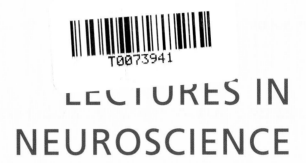

LECTURES IN NEUROSCIENCE

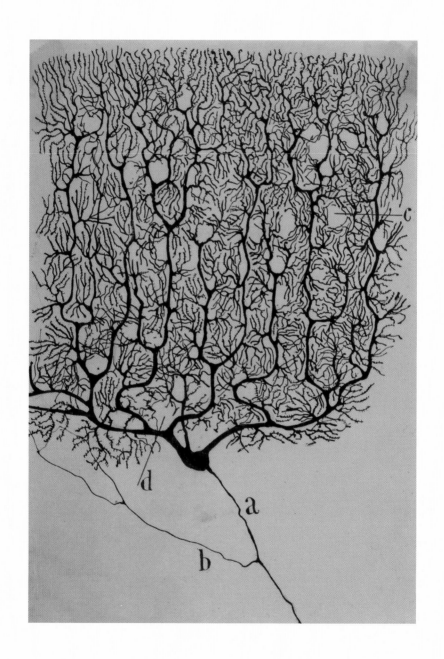

LECTURES IN NEUROSCIENCE

RAFAEL YUSTE

COLUMBIA UNIVERSITY PRESS | NEW YORK

Columbia University Press
Publishers Since 1893
New York   Chichester, West Sussex
cup.columbia.edu
Copyright © 2023 Columbia University Press

Illustrations copyright © 2023 Gil Costa
All rights reserved

Library of Congress Cataloging-in-Publication Data
Names: Yuste, Rafael, author.
Title: Lectures in neuroscience / Rafael Yuste.
Description: New York : Columbia University Press, [2023] |
    Includes index.
Identifiers: LCCN 2022041320 (print) | LCCN 2022041321 (ebook) |
    ISBN 9780231186469 (hardback; acid-free paper) | ISBN 9780231186476
    (trade paperback; acid-free paper) | ISBN 9780231546652 (ebook)
Subjects: LCSH: Neurosciences—Textbooks.
Classification: LCC RC341 .Y87 2023 (print) | LCC RC341 (ebook) |
    DDC 612.8—dc23/eng/20220923
LC record available at https://lccn.loc.gov/2022041320
LC ebook record available at https://lccn.loc.gov/2022041321

Columbia University Press books are printed on permanent
    and durable acid-free paper.
Printed in the United States of America

Cover design: Lisa Hamm
Cover image: Neuronal circuits in human cerebral cortex. Courtesy of the
    Cajal Institute, "Cajal Legacy," Spanish National Research Council
    (CSIC), Madrid, Spain.

Frontispiece: Purkinje cell from human cerebellum. Courtesy of the Cajal
    Institute, "Cajal Legacy," Spanish National Research Council (CSIC),
    Madrid, Spain.

For Sarah Faigela, Mindl Esther, and Chaya Doveva

CONTENTS

PREFACE

The purpose of this book is to answer, in simple language, two questions: what the brain does and how it does it. The brain is one of the most fascinating objects in the universe, a few pounds of squishy matter that builds our mental world. Thus, if we understood the brain, we would understand our minds, who we are, the center of our human experience. As humans, we are defined by our minds and we have struggled to understand ourselves, attempted to do throughout our history as a species. This struggle has culminated in efforts to understand how the brain functions. For these reasons, many scientists—myself included— think that neuroscience is the most fascinating topic to which you can devote your life.

But despite our efforts, we still don't have a general theory that explains how the brain works or generates our mind. This is likely due to methodological limitations. Neuroscientists have traditionally tried to decipher the function of the brain by studying how neurons work using methods to record the activity of individual neurons, taking the brain apart one neuron at a time to put it all together at the end. But the human brain has close to 100 billion neurons, and, lacking methods to measure the activity of many neurons operating at the same time, we have been missing the bigger picture, how neurons interact as a group. Studying individual neurons one by one makes it impossible to reveal the brain's function, the same way that looking at individual pixels on a TV screen will never enable us to see images and understand the function of the TV or the meaning of the movie playing on it. In my view, neurons acting together are like the pixels that form a particular image by turning on at the same time. These groups of neurons, or "ensembles," are the functional unit of the brain, the images of our "brain TV."

That viewpoint is my focus in this book: that ensembles of neurons form the building blocks of brain function and that the brain uses these groups of neurons acting together to build "emergent" functional properties. Emergent properties, like magnetism, are properties of systems of many units (atoms in the case of magnetism) that are absent from the individual units themselves. Thus, by definition, these properties cannot be understood by investigating individual elements one at a time; they appear only when elements interact. In fact, one could argue that the brain is the mother of all emergent systems as it is built with an astronomical number of neurons connected by an even larger number of connections. If you wanted to build a system to generate emergent properties, you couldn't do better than invent the human brain.

The study of emergent properties is at the frontier of science, as interactions between units in complex systems are difficult to capture experimentally or mathematically. Nevertheless, emergent properties are all around us. They dominate our world, from subatomic particles building atomic properties, to atomic interactions building molecular properties, to molecules building chemical properties of matter, to chemical properties building biological structures, to biological structures building living matter and mental activities. Nature is a ladder of emergent properties.

This book will focus on one particular emergent level, that of "neural circuits"; that is, groups of connected neurons. To me, understanding how the brain works means understanding how neural circuits carry out computations and mental processes. You could argue that how neuronal activity is transformed into thought is the key question in neuroscience. And we are aiming for that heart of the matter here. By concentrating on neural circuits, this book goes against the grain of many neuroscience textbooks that for the last two decades have focused, with increasing reductionism and exquisite detail, on molecular and cellular neuroscience. There are also textbooks that analyze the output of the brain at a behavioral, ethological, or psychological level. But the argument I make is that the heart of brain function lies in the middle of the ladder, at the circuit level, connecting neuronal activity to thoughts via neuronal ensembles. Molecules and cells are, of course, critically important, but I don't think they will reveal the basic principles of how the brain works. And going from behavior to understanding how the brain works, though it can reveal key mental computations, risks treating the brain as a black box, which may prevent understanding of the basic principles of its function.

My intuition tells me that the critical insights lie inside the box, in the middle of the apparently complicated neural circuits formed by myriad neurons and their connections, the "impenetrable jungles where many investigators have lost themselves," as Santiago Ramón y Cajal, whose beautiful drawings and work have inspired neuroscientists like me for over a century, sternly warned us. But in contrast to Cajal's opinion that brain circuits are impenetrable, my argument—in line with that of many previous investigators, such as Thomas Graham Brown, Rafael Lorente de Nó, Donald Hebb, Rodolfo Llinás, John Hopfield, Valentino Braitenberg, and Moshe Abeles, among others—is that groups of neurons acting together are basic modules that simplify our understanding of brain function. To understand the jungle, we need to look at it not tree by tree but from above, and observe how groups of trees make up the forest.

This book arose from class notes and blackboard sketches for my undergraduate course, Neurobiology II: Development and Systems Neuroscience, which I have taught at Columbia University since 1996. Year after year, I have struggled to present a coherent picture of how the brain works to groups of fresh, smart, and sharp-minded undergraduate and graduate students. After many years of using textbooks that keep getting longer and heavier and bury us in a molecular tsunami of exquisite but ever more complex details, I thought it was time to step up to the plate, try to go back to the basic principles, and have a go at writing a textbook with a synthetic angle. So here it is.

This book therefore is based on my Columbia University lectures and is meant for newcomers to neuroscience. It should be accessible to laypeople, as it was written for undergraduate students with no previous background in neuroscience. It is not a comprehensive review but a personal synthesis of the field. In eighteen lectures, I will touch on examples to review and extract basic principles. For those interested in a deeper dive into details or comprehensive treatment of molecular, cellular, behavioral, or computational neuroscience, I recommend the excellent textbooks already published, which are phenomenal pieces of work and scholarship. My goal here is not to compete with them and add yet another textbook to this long list of excellent volumes. Rather, I aim to bring back the essence of brain function using a neural circuits framework to reduce and synthesize important information from the larger textbooks into a short, succinct text you can carry in your pocket. My mission is to capture the excitement that brought me to neuroscience in the first place.

I have one caveat for this journey of ours: by design, to make a succinct synthesis, I must leave out a lot of details. The goal is to present a clear picture, and for the purpose of this introductory text, I value simplicity and clarity at the expense of detail and exhaustiveness. Let's not get lost in that jungle. I want to be didactic and review a few examples as threads in a story, repeating key concepts on purpose. Also, instead of complex illustrations, simple drawings sketch the essence of an idea. I am reminded of a comment from my mentor, Sydney Brenner. He said, with a mischievous smile, that when you're trying to understand something, you can choose to ignore some facts that may get in the way, as they will eventually be explained sooner or later.

Also, while I cover many parts of the brain, the thrust of the argument I will lay out has to do with the cerebral cortex, which is my area of expertise. The cortex (meaning "bark" in Latin) is a thin sheet of tissue, about 2 mm thick and a meter square in area, that covers the top of our brain. The cortex is the largest part of the nervous system in mammals and is important because it generates all our mental and cognitive abilities. Our extraordinary cognition is possible because of its huge size; it is much larger, in proportion to our body size, than that of any other animal. The cortex is the reason I am interested in neuroscience. We need to understand all of the brain, but the cortex is where it's at for me.

The book would not have been possible without the help of many people. Gil Costa, a neuroscientist and artist, illustrated every concept with beautiful figures and provided artistic and scientific feedback.

Lisa Hamm piloted the art design with aplomb and gusto, while Miranda Martin skillfully edited the book and shepherded us all to the finish line. My former undergraduate student Polina Porotskaya transcribed many of the lectures and also sketched initial versions of many illustrations. Jay Walkers, another former student, helped transcribe the motor system lectures. And Ricardo Martínez Murillo, Director of the Cajal Institute, provided access and permission to reprint the beautiful drawings of the master.

I also acknowledge my friend Sebastian Seung for cowriting a previous piece that I used for inspiration for the neural network lecture. Also, heartfelt thanks to my colleagues at DIPC in San Sebastian for their comments, and to my dear colleague Pat Kitcher, who shared her passion about Kant, fueling mine. Most important, Columbia University and particularly its Department of Biological Sciences provided a supportive, warm, and exciting home from which to work, think, and teach. But I am most thankful for twenty-six (and counting!) generations of undergraduate and graduate students who, with their questions, insights, and fresh spirits, sculpted my thinking and lectures over the years. And finally, to my family—the *etxekoak* (the people from the house, as they say in Basque)—whose love, patience, and support has provided me the roots from which to grow.

LECTURES IN NEUROSCIENCE

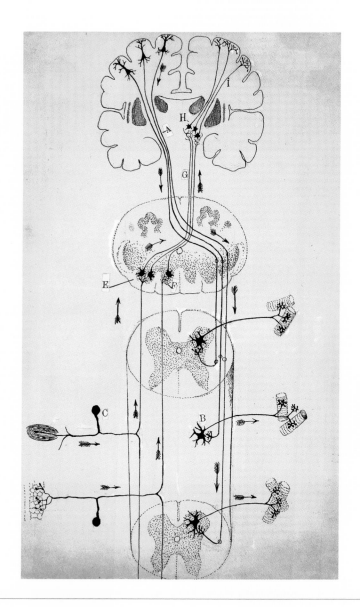

The brain as a sensory-motor machine. The mammalian nervous system was originally viewed as a reflexive machine that takes sensory inputs, and processes them in the cortex to generate an output. Courtesy of the Cajal Institute, "Cajal Legacy," Spanish National Research Council (CSIC), Madrid, Spain.

LECTURE 1: PRINCIPLES

The great topmost sheet of the mass . . . becomes now a sparkling field of rhythmic flashing points with trains of traveling sparks hurrying hither and thither. The brain is waking and with it the mind is returning. It is as if the Milky Way entered upon some cosmic dance. Swiftly the head mass becomes an enchanted loom where millions of flashing shuttles weave a dissolving pattern, always a meaningful pattern though never an abiding one; a shifting harmony of subpatterns.

—C. S. Sherrington, *Man on His Nature* (1942)

OVERVIEW

In this lecture, we'll learn

- ▶ What the brain does
- ▶ How the brain does it
- ▶ The basic principles of neuroscience

What the Brain Does: Predict the Future in Order to Act

I would argue that the purpose of the brain is to predict the future so the animal can act successfully. Prediction–action—that's the brain in two words. To predict what will happen, the nervous system builds a model of the world, more or less complex depending on the animal, mentally internalizing the physical world so it can be run as a kind of virtual reality model of the future. This virtual reality model is exquisitely built with maps inside our brain and updated internally using memories from our life's experiences and kernels of ancient biological wisdom ingrained by evolution in our nervous system. Using this information, and continuously updating the accuracy of the model with our senses, we anticipate the future and program our actions so they can help ensure our survival.

Predicting successfully is the heart of being intelligent: to guess what the future will bring and act accordingly so you are on top of the situation and can avoid the problem—or, if you can't avoid it, at least be ahead of the competition. As the physicist Leo Szilard put it, "In order to succeed, it is not necessary to be much cleverer than other people. All you have to be is one day ahead of them." The ability to predict must be a fundamental advantage for survival in evolution; no wonder brain-laden animals dominate the earth, culminating with the brainiest of them all, *Homo sapiens*.

The idea that the brain builds a model of the world is not new, echoing back to the birth of idealism in Plato's cave with the idea that the world is different from our ideas of it. This argument was made particularly clearly in 1781 by Immanuel Kant in his *Critique of the Pure Reason*: the mind constructs representations (*Darstellungen*) of the world, and we live on an island of these representations, isolated in an ocean of reality that we do not experience directly. Many people since Kant, including Hermann von Helmholtz, Sigmund Freud, and Kenneth Craik, have refined the idea that internal states of the brain represent the world, proposing that we use these representations to build a model of the world, which we manipulate mentally instead of manipulating the physical world. We fast forward the tape, so to speak, to predict what happens and act accordingly. That is the prediction–action theory of brain function, which, as the name says, is still a theory.

So why do we say that the purpose of the brain is to predict? An important hint comes from evolution. One way to figure out what the brain does is to look at species that have nervous systems and compare them with species that don't (figure 1.1). The only species with a nervous system are essentially those that move. The nervous system is associated with the ability to move. About 750 million years ago, the first animals with neurons appeared in evolution, right before the split between cnidarians, which are radially symmetric species like jellyfishes, sea anemone, and corals, and bilaterians, which, like us, are animals with bilateral symmetry.

Why animals have radial or bilateral symmetry is a fascinating question for another time. Before these two phyla split, animals from older phyla likely had no neurons, and their descendants today—like poriferans (sponges)—stay attached to the same rock their entire lives. Meanwhile, cnidarians could swim around, explore their environment, and, most important, hunt. And bilaterians could not only also do those things but were able to take over the seas, the land, and the air.

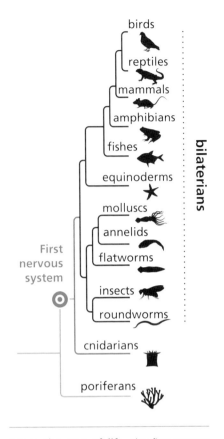

1.1 In **the tree of life**, the first nervous systems appeared before the branching of cnidarians and bilaterians.

So why would you need a brain if you move? If you're moving, you need to be able to predict the future. That is critical. You're going to be somewhere else, and need to know what and who is around you (prey, predator, mate, for example) and what to do in this new environment. Otherwise your behavior may not fit the new environment and could quickly lead to your demise. Moving is risky business. But importantly, by moving, you're propelling yourself not just in space but also in time, toward a different future, which you have crafted by your actions. Movement means future.

So how does one predict the future? First, you need an idea of what you expect will happen. This idea could be generated by an internal model of your world based on your memories and past experiences. This model has to be located somewhere, and it makes sense that it be organized into maps, so you can find your way in it. You then extrapolate this model into the future by mentally manipulating its parts—the Kantian "representations" of your body and environment—and compute how the future should look.

After prediction comes action: you make a decision, you move your muscles, you act. By moving, you physically enter the future with your body and perceive what happens, what's out there. You then measure this new state in which you find yourself with your senses, turning the future into present. The cycle ends when you compare your assessment of what happened with your internal model, measuring the error in your prediction, and then refine the model so you get it right the next time. This updating is called *learning*. You add to or remove things from the model, fine-tune it up or down so you are ready to do a better job next time, since your survival, or reproduction, depends on outperforming your competitors or outsmarting your predator or your prey. Learning through sensory experience improves the model accuracy, to the point that it may not be surprising that one could confuse the model with reality.

This view of the brain as predictor for action selection may seem logical and straightforward, but it is not widely accepted. In fact, the traditional way of thinking about brain function is that it is a sensory-motor system, an input-output machine. This view, anchored in the way reflexes work, considers that the purpose of the brain is to generate behavior—its output—in response to a sensory stimulus—its input. From this perspective, the brain works like a machine: in comes the input, some gears kick in, and out goes the output. This reflex action view of the nervous system, beautifully illustrated by Cajal in the drawing that starts this lecture, is also logical and earned its proponent, Charles Sherrington, a Nobel Prize. This sensory-to-motor view has infused neuroscience with a research program that aims to map these reflexes, all the neuronal cranks and levers by which sensory stimuli generate behavior. Stimuli give rise to perception, which triggers decision, which in turn triggers behavior. The brain would be a gigantic phone book, or a look-up table in a computer program, in which every input corresponds to an output. To decipher the brain, our job as neuroscientists would be simply to fill in this phone book line by line so you could then read the function of the brain like an instruction manual for a manmade device.

The prediction–action view differs importantly from this sensory-motor machine perspective. If the brain is a predictor, the key part of its workings is its internal program, not its reaction to sensory inputs; it's no longer a reflex. In fact, you can run the internal program without inputs, as we will see in the following lectures, and choose whether or not to generate behavior. Our virtual reality machine can precisely couple itself to the world, as when we are awake and attentive, or decouple itself, as when we are daydreaming or sleeping. Indeed, in our dreams, we are probably operating our same virtual reality machine in a free-running mode, disconnected from reality. As the Spanish playwright Pedro Calderón de la Barca, put it: "Life is a Dream."

Dreams are hard to explain with an input-output model. In fact, the existence of any spontaneous activity in the brain—that is, internal activity that is not triggered by the senses—was the origin of the first shot fired at the Sherrington reflex model, a rebellion that, interestingly, started in his own lab. It was his own student, Thomas Graham Brown, who discovered endogenous patterns of neuronal activity generated by interconnected groups of coactive neurons and called them *central pattern generators*, or CPGs. Brown argued that CPGs, not reflexes, generate behavior and that using them, the spinal cord could intrinsically generate walking even without sensory inputs.

Today, this way of thinking is represented by models of predictive coding and processing proposed by Karl Friston and many others. These models often use Bayes theory, as calculating probabilities of past events is the best statistical way to predict the future, and formalize these ideas in a mathematical framework to enable the design of experiments to test whether or not the ideas can be validated. Different models use different nomenclatures—like surprise minimization, free energy, and reinforcement learning—and have important differences, but they essentially agree that animals generate statistical models of their environments to accurately predict the future, minimizing surprising or life-threatening conditions. To do this effectively, animals must also generate an internal predictive model of themselves in relation to their environment and change it to minimize their model's long-term surprise, or short-term prediction error. In short, they need a way to learn from the outside world to continuously update the model and statistically maximize their model's accuracy. Indeed, as we will see in later lectures, the brain is exquisitely sensitive to the occurrence of unpredicted (novel) stimulus or to the nonoccurrence of predicted ones. The detection of novel stimuli even happens during anesthesia!

This non-Sherringtonian way of thinking thus focuses not on how the brain responds to the world reflexively but on how the predictions about reality generated by intrinsic brain states are compared with incoming sensory information. If we think in those terms, brain evolution makes perfect sense. The original purpose of the nervous system—to enable movement—appears in the simplest invertebrates and becomes more sophisticated over evolutionary time. CPGs, generated internally by neural circuits initially associated with movement or behavioral sequences of them in fixed action patterns, which orchestrate motor patterns such as swimming, eventually become symbolic and encode abstract concepts rather than muscle patterns. This "encephalization" of CPGs takes over; the forebrain outgrows the older spinal cord and movement circuits and, in the case of the cerebral cortex, becomes a sort of universal computer, a biological machine that encodes and mentally manipulates the world using neuronal ensembles as symbols. In the case of humans, it generates our mind. In fact, it generates our world. Echoing Kant, we may think we live in a real world, but the reality we perceive is likely entirely constructed or generated internally. Our world is the program running in a prediction machine, and it's such a good model that we interpret it as reality.

In this evolutionary game, to predict better and better, as you need a model of your body, a critical step must have been the invention of the self. The "I," the concept of "me," could be the ultimate invention of evolution, because it enables us to include our own body in this virtual reality model and mentally manipulate ourselves in relation to the world. In fact, the behavior of an animal does not make sense unless it is anchored in the concept of a unique self. This may seem like the silliest question, but how do you know that you are one person and not many? How does our nervous system know that the behavior that it is about to generate applies solely to our body and is consistent with our past and present? There has to be some way for the nervous system to generate the idea of me, of self, of a single unity of a behavioral program and stamp it in all that it does. And that could apply not only to humans but to all species that have a nervous system. Otherwise, behavior would not be coherent or consistent, and those species would not survive the evolutionary arms race to eat, not be eaten, and reproduce. How this conceptual unity, or self or consciousness, is generated is one of the key questions in neuroscience.

A final thought on species without brains: bacteria, unicellular animals, and placozoans, for example, can move, find food, and mate yet lack nervous systems. Shouldn't this kill our hypothesis? Well, there is a significant size difference. If you are small or have only a single cell, you don't really need neurons to move (you don't even have space for them). But once you grow to a certain size, you need to coordinate different parts of your body for coherent, effective behavior. Although it is difficult to discern what exactly happened 750 million years ago, the magical time when the first nervous system appeared in evolution, it seems reasonable to assume that neurons were first needed to coordinate activity across many muscle cells once the body became too large and developed many cells located too far apart to communicate among themselves. The nervous system, with its fast impulses traveling along long wires, can do that job and enabled multicellular animals to grow larger in size, giving them a leg up in the competition.

How the Brain Works: Principles of Neuroscience

We now have an answer to the "what is the goal of the brain" question: to predict the future in order to act intelligently. Let's now deal with the "how" question: how does the brain do all this? What are the biological mechanisms, or tricks, that evolution has invented to do this job? How is this virtual reality program created, and where does it live? The simple answer to all these questions is through *neural circuits*; that is, neurons working together. These circuits form the brick and mortar of our models of the world and are updated with experience. Although how exactly this happens is not yet clearly understood, we can start to discern some general rules, some principles. These are the mechanistic principles of brain function, the nuts and bolts of the machine, and they are intertwined, sort of built up on top of one another. We can synthesize and summarize them in seven principles of neuroscience (box 1.1). That is the gist of the course, so pay close attention. We are going to walk through all seven principles one by one and explore how they are interrelated.

BOX 1.1: Principles of neuroscience

- ✓ Ensembles
- ✓ Hierarchy
- ✓ Wiring
- ✓ Learning
- ✓ Maps
- ✓ Control
- ✓ Optimization

Neuronal Ensembles

Let's unpack this neural circuit idea. The critical building block for mentally manipulating reality is the concept of a thought. A thought can be anything mental: a perception, memory, idea, movement plan, emotion, and so on. But in reality, a thought is just a symbol of something, a representation, as Kant said. And how do you build such representations in the brain? One way is by having neurons activate in groups, as an ensemble (figure 1.2), firing together during a brief moment of time, the duration of the thought. So that is our first principle of neuroscience: that

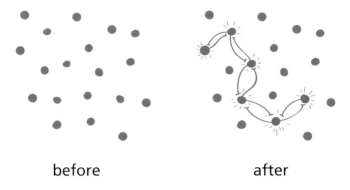

before   **after**

1.2 A **neuronal ensemble** is a group of neurons that, after being stimulated, connect to each other and become active as a unit, building an internal brain state.

neuronal *ensembles*—that is, groups of neurons firing together—build internal symbolic structures.

Importantly, neurons inside these ensembles connect to one another and, by doing so, activate themselves and keep the ensemble on, keeping the ball in the air, so to speak. This way, thoughts can exist internally without an outside stimulus. You need to be able to do this if you want to manipulate thoughts internally. In order to think, you need to break from the tyranny of the outside world, distance yourself from the senses, and keep an idea active in your mind. Ensembles are found in the cerebral cortex and in many other areas of the brain. In the fourth lecture, we will see how their joint firing has been linked to specific behaviors or mental states.

Ensembles could be the "images" that are playing on the brain "TV." And the purpose of the cortex, and all this apparently complicated neural hardware, could be simply to build a large assortment of ensembles that, as internal states, could represent the world as symbols. Just as humans generate language with words, the cortex builds a symbolic language through the firing within ensembles of neurons. In a mysterious transformation, the combination of ensembles—their grammar—somehow generates our minds and those of animals in the same way that you generate a story with words. Thus, neural circuits, these impenetrable jungles, are vast and fascinating structures that build a multitude of fleeting flashes of activity, as Sherrington imagined, with each ensemble representing a particular concept. This endless dance of thoughts, the discourse of our world, constitutes our minds and only ceases when we die.

Hierarchy

Our second principle is that these ensembles are not just dancing around without rhyme or reason but are organized in *hierarchies*. Ensembles are modules; that is, a set of independent units that can be used to construct a more complex structure. Just like a Lego set, there is an infinite number of possibilities, and the sky is the limit as to what you can build with them. Modularity also enables compositionality; that is, the possibility of switching modules in and out. Modularity also allows ensembles to be organized in levels; for example, ensembles at a top level could become activated by a combination of lower-level ensembles and represent them as a symbol of lower-level symbols. You could then build more abstract symbolic meaning as you go up in level. In evolution, the more complex the brain becomes, the more levels it has, and the hierarchies get larger and larger. That's why ensembles are so powerful. A system of them doesn't need to be made up of an enormous number; they can be combined to represent the world and do so at different levels of abstraction. Just like human language.

Wiring

How do you build these modules and link them into hierarchies? You do so by *wiring* neurons together in particular ways. That is our third principle of neuroscience: neural circuits and hierarchies of ensembles are connected through a *specific wiring diagram*. This wiring diagram mathematically implements computational algorithms. Neural circuits throughout the brain are composed of many hundreds of cell types, which are connected in patterns that we are only now starting to under-stand. In fact, how neurons are wired remains one of the open questions in neuroscience. Though we remain somewhat ignorant about the pro-cess, we do know that to achieve an organized and effective system of ensembles, wiring must be precise. There are different ways to achieve a specific connectivity, and the brain uses two strategies, or two design principles of wiring diagrams, both present even in the same parts of the brain of the same animal. Sometimes it pays to be specific, and sometimes it's better to be distributed.

In the first type of wiring diagram, connections are precise from the start and are built with great care, as we will see in the next lecture. In a nutshell: precision in, precision out. This is also known as the *labeled lines* design, in which neurons choose to connect only with other neu-rons, and their connections carry specific information. For example, a group of neurons are coding for sweetness and send this information to other neurons that care only about sweetness. That way, you can map anything in the world into small drawers in your brain and use that information later as you see fit. That makes perfect sense if you are building a model of the world. Labeled lines are also great for building reflexes and mapping sensory stimuli into motor actions, which by itself can explain the sensory-motor model of the brain. As we will see in this book, there is a lot of evidence for a labeled lines design; for example, in the spinal cord and in many parts of the brain. But there is more to the brain than labeled lines.

The second wiring diagram involves starting with a distributed connectivity—that is, where a neuron is indiscriminately connected to many other neurons with no specificity at the beginning, but then the connections get pruned or tweaked up and down through learning and experience, so you end up with a selective neural circuit. In other words, random goes in, then learning, then out comes precision.

You craft selectivity out of a primeval soup. Indeed, there is evidence that neurons in many parts of the brain seem to send axons indiscrimi-nately to many other neurons, and vice versa: many neurons receive con-nections from many other neurons. This is what Santiago Ramón y Cajal called the "telegraph poles" type of wiring, whereby axons seem to be touching everything in their path. Ideally, each neuron could be connected to all the other neurons and receive connection from all of them. That would constitute a completely distributed circuit, and, in cases like what occurs in the cerebellum, nature seems to be trying to achieve this goal. This distributed wiring doesn't need to be complete; in fact, it is okay if it is random, meaning that each neuron could be connected to a random assortment of other neurons, as could be happening in parts of our cerebral cortex. This is because mathematically, a random connectivity is equivalent to a completely connected circuit.

There are many things you can do with distributed circuits that you cannot do with labeled lines, as we will see later in the book. But there is a price to pay with distributed wiring: the more neurons you are connected to, the less selective you are. And, if you receive information from all neurons, it becomes difficult to be specific. In fact, it's the exact opposite of specificity, as a given neuron could be involved in many computations. But didn't we say that wiring is specific?

Learning

This is where our fourth principle comes to the rescue: *learning*. A variety of mechanisms during the life of the animal shape this distributed connectivity, pruning the connections that are not needed, making some connections stronger and others weaker until you achieve the specific circuit you need. This is similar to what electrical engineers do in field-programmable gate arrays (FPGAs), electronic circuits in which everything is connected to everything else and which is then modified. If you are trying to design an electrical circuit for a new job but are not sure how to go about it, you begin with a master FPGA, in which every gate is connected to every other gate, and in the field you carve away the connections that you don't need. FPGAs are thus multipurpose: you can use them to do any job by pruning them. That's how you turn distributed wiring into something useful. And, like FPGAs, the brain—or at least some parts of it, like the cortex, hippocampus, and cerebellum—seem to be multipurpose, involved in all kinds of computations.

In fact, the entire brain is essentially a learning machine, always drawing information from the world and storing it. Learning is a key component of the prediction–action model, as it enables us to improve our predictions based on the result of our actions and build better models, which often happens by associating new stimuli or concepts. For example, you move to a new neighborhood in Paris and have to painfully learn that the bakery closes on Mondays, as happened to me. So you associate the concepts "Monday" with "zero bread" and use that to compute a new behavior, buying extra bread on Sundays or buying bread outside the neighborhood.

Associative learning is one of the cornerstones of psychology, and there's ample evidence that many parts of the brain are engaged in it. One of the ways this associative learning could occur in neural circuits is captured by Hebb's rule, proposed by the Canadian psychologist Donald Hebb in 1946, which states, as Carla Shatz famously summarized it, "neurons that fire together, wire together." Imagine that you have two neurons, each responding to a particular stimulus. Now picture both stimuli happening at the same time over and over again. Our goal is to build an ensemble in which these neurons will start working together as a unit, which means that every time that you receive one stimulus, you activate the neurons responsible for the second one. Through Hebbian synaptic plasticity, you strengthen the connections, and the action becomes automatic, and you encode the causal relation between both events.

There are other, non-Hebbian, ways that this occurs, but the end result is the same: you learn to associate both stimuli and build a new concept, a new representation. Similarly, you can gather ensembles together into hierarchies by linking particular neurons from each ensemble, ones that can bring the entire ensemble along, as we will see in the coming lectures. Finally, if we have a distributed circuit—let's say with everyone connected with everyone else—you could in principle link any neuron and ensemble, any concept. There are no limits to the mind.

Maps

We have now examined ensembles, hierarchies, wiring, and learning and seen how these four principles of neuroscience beautifully interlock. It gets even more beautiful, because it turns out that specific wiring diagrams are neatly organized into *maps*, which is our fifth principle.

The brains of all animals are full of spatial topographic maps. These maps can represent features of the sensory world, like the position in space of a visual stimulus (captured by a corresponding position in the retina or retinotopy). Other maps represent auditory frequency, somatosensory inputs from different parts of the body, or motor maps, where the synergistic coordination of muscles is organized. Some of the most interesting maps are cognitive or computational ones, in which the feature being ordered is an abstraction, calculated by the nervous system itself, like depth perception, color, movement, and abstract patterns, which we are only beginning to understand. We probably also have maps of concepts, words, emotions, future events, and more.

Whichever variable the brain analyzes is mapped in space in an orderly fashion. In fact, Kant would likely argue that all maps are cognitive, even sensory maps. According to this view, the maps of the visual world in our brains do not really represent the world; rather, they represent our *idea* of the world. It may seem like a small difference, but it's a fundamental one. In fact, in the lecture about vision, we will see concept at work when we encounter the retinal blind spot, a region of our visual map that does not receive any visual information but which gets filled in by our brain. And if this recreation of the sensory reality by the brain happens there, it probably happens everywhere.

Why are brain maps important? Maps are actually models, representations. They *are* the famous model of the world we have been talking about. A map of the world is a model of the world. We don't have just a single map but many, and each is a slice of the mental world we live in. These maps are linked and interact in precise ways. For example, in our superior colliculus are maps of the auditory space around us, precisely aligned with maps of the visual space.

But why use physical maps at all? Couldn't you just wire ensembles as they come without having to place them side by side in order? While some people think that maps are developmental epiphenomena with no real function, the orderly organization of information in space seems important to enabling fast and efficient processing and computations. It is also a beautiful demonstration that evolution can be logical and systematic, qualities that probably go hand in hand with enhanced survival and economic use of energy resources. Evolutionarily, order is good.

This hierarchical architecture of maps could be the framework of our minds and of our world and is likely reflected in the physical structure of brain areas and their connectivity. So when you look at a picture of a brain or of its connectivity, you may be looking at the architectural plans of the mind.

Control

Now that we have an idea of how the brain hardware is organized with our five initial principles, let's circle back to how you use all the hardware to predict the future and choose your action. Let's talk about the software that is running on the brain's hardware. Our sixth principle of neuroscience is *control theory*, which explains how the brain can predict, act effectively, and update those predictions based on the outcome.

What exactly is control theory? It is a mathematical way by which you compare an output (say, a motor action) with a prediction, an expectation of what will happen, and use the comparison to compute an error signal, then use that error to fine-tune your output so it becomes more accurate in the future. It's a feedback system, similar to a thermostat, that adjusts the next step up or down. In fact, feedback control is the way most human-built engineering systems work. Besides thermostats, mechanisms for steering ships, elevators, washing machines, cars, planes, electronic circuits, and computer programs are designed following principles of control theory to make these machines run more smoothly. One could argue that the brain is a gigantic control system. As neuroscientists, we therefore must often put on our engineering hats and use those tools and approaches to reverse-engineer these circuits and figure them out.

The exact mechanisms of how feedback control occurs in the brain are an exciting area of current research. Different strategies seem to be in play in different brain areas. For example, it seems that many parts of the brain are exquisitely built to detect novel stimuli. Novelty detection is a must if you want to measure sensory input and match it to your expectations to adjust them. In other words, novelty detection *is* the error signal of a control system, when you experience something that you did not predict. How could this novelty detection signal be used to adjust our internal models? Using reinforcement learning, an error signal—this novelty detection signal—could be fed back into the system so it can strengthen certain synapses at the expense of others. Dopamine seems to be playing a big role in reinforcement learning, which appears to be also one of the most powerful ways to learn in neural networks.

Besides novelty detection and reinforcement learning, another strategy that seems to be at play is the computation of Bayesian probabilities. Apparently, in both the behavioral response of people or animals and the activity of certain neurons, the brain seems to be faithfully computing probabilities of events happening. And if you are in the game of statistics, following Bayes theorem is, mathematically, the best you can do in terms of calculating the probability that something will happen in the future. That is why some people think the brain is essentially a Bayesian machine, continuously learning and updating a statistical model of the world.

Optimization

Finishing our engineering job, here's our seventh and final principle of neuroscience: *optimization*, meaning doing the job effectively and with minimal cost. To carry out this entire prediction–action task effectively without wasting energy and time, the nervous system harnesses the physical constraints of its own biological hardware. To do that, neurons compute using the physics and chemistry that they are made of. This is as effective as its gets: you use the material you are built with to compute. No extra weight in your backpack! In other words, neurons turn every aspect of their biology into computational elements.

In this respect, the brain is not at all a digital computer; it's a biological one, an organic computer that is anchored in the physics and chemistry of its hardware. It uses its biological hardware as a computational device. The purpose is efficiency, or optimization. Indeed, the brain is an extremely efficient machine, using hardly any power, equivalent to about a twenty-watt light bulb, yet it can compute problems that even the largest supercomputers cannot. The number of connections inside a human brain is estimated to be three times the nodes of the existing internet on the earth. And this gets powered with twenty watts! That ability is obviously something evolution has selected for, since the less power you need and the more efficient you are, the less food you need to eat and the longer you can last without food, outcompeting your competitors. Evolution has been working on improving the energetics and optimizing the structure and function of the brain for almost a billion years. As we decipher how different parts of the brain work, we are astounded every step of the way by the elegance and efficiency of the mechanisms underlying their computation.

In the case of the sensory systems, which are better understood than other parts of the brain, it seems that this efficient design has pushed the hardware to the physical limit of the natural world, so that retinas can detect individual photons, ears can listen to the sound of individual atoms moving, and olfactory systems can detect the appearance in the air of individual molecules. It doesn't get more sensitive than this. Sensory systems have been optimized for the physical limit, so it makes sense to expect that the algorithms and neural hardware that are handling sensory information must be equally optimized, perhaps also to the mathematical limit. The brain is a learning machine that stores the statistics of the world, predicts the future, can explore an infinite number of possible computations, and, on top of that, is a machine that runs essentially on a sandwich and a little bit of water!

▼ RECAP

One could argue that the brain does just one thing: predict the future using control algorithms to act intelligently. A few common principles of neuroscience explain how this is done beautifully and efficiently using neural hardware. A model of the world is built with neuronal ensembles that are symbols of things. Animals use these ensembles in all kinds of neural circuits for, effortlessly computing optimal strategies for future behavior from an enormous number of possibilities. They implement intelligence. What an extraordinary job. Let's dive into it. Let's find out how the brain is doing this. But first, in the next lecture, let's take a closer look at the structures that make up the brain and how they develop. What exactly *is* the brain, what does it look like, and how does it assemble itself in the first place?

Further Reading

Craig, K. J. W. 1943. *The Nature of Explanation*. Cambridge: Cambridge University Press. One of the earlier proposals that brain functions by the manipulation of internal representations of the world.

Friston, K. 2010. "The Free-Energy Principle: A Unified Brain Theory?" *Nature Reviews Neuroscience* 11, no. 2, 127–38. This review, likely to become a classic, describes the idea of the brain as a prediction machine.

Kandel, E. R., J. D. Koester, S. H. Mack, and S. A. Siegelbaum. 2021. *Principles of Neuroscience*, 6th ed. New York: McGraw Hill. A well-written and even better-illustrated encyclopedic reference of neuroscience. The "bible." An excellent background reference for most of the lectures, edited by a world-class team.

Llinás, R. 2002. *I of the Vortex: From Neurons to Self*. Cambridge, Mass.: MIT Press. A very lucid description of the importance of internal states for brain function. One of my favorite books.

Luo, L. 2020. *Principles of Neurobiology*. New York: Garland Science. Coherent coverage of modern neuroscience by one of the leaders in molecular neuroscience.

Purves, D., et al. 2017. *Neuroscience*, 6th ed. Oxford: Sinauer/Oxford University Press. A comprehensive yet approachable introduction to neuroscience. Beautifully written and edited.

Ramón y Cajal, S. 1899/1995. *A Histology of the Nervous System of Man and Vertebrates*, repr. Oxford: Oxford University Press. A classic, and beautiful, description of the structure of the nervous system. Cajal essentially nailed it. If you could save only one neuroscience book for future generations, this is it. Following the tradition from my Ph.D. mentor, doctoral students in my lab receive a copy of this two-volume set when they graduate.

Ramón y Cajal, S. 1917/2004. *Advice for a Young Investigator*, repr. New York: Bradford. A refreshingly candid one-on-one with the master. Cajal himself gives you advice about what to do with your life. Precious.

Shepherd, G. M. 2003. *Synaptic Organization of the Brain*. Oxford: Oxford University Press. A classic description of the principles of brain circuitry.

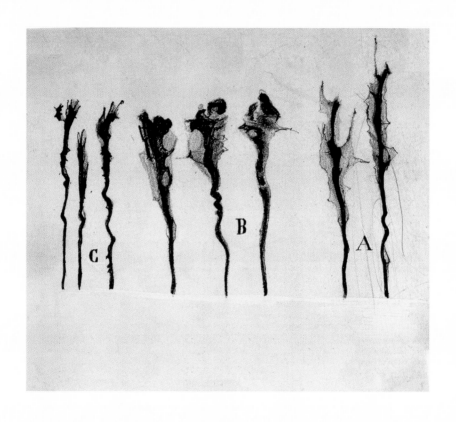

Neuronal growth cones in the spinal cord of a chick embryo. These hand-like structures, on the tips of developing axons, carry out the most amazing feat of neural development: its wiring. Courtesy of the Cajal Institute, "Cajal Legacy," Spanish National Research Council (CSIC), Madrid, Spain.

LECTURE 2: BRAINS

After a short period spent in Brussels as a guest of a neurological institute, I returned to Turin on the verge of the invasion of Belgium by the German army, Spring 1940, to join my family. The two alternatives left then to us were, either to emigrate to the United States, or to pursue some activity that needed neither support nor connection with the outside Aryan world where we lived. My family chose this second alternative. I then decided to build a small research unit at home and installed it in my bedroom. My inspiration was a 1934 article by Viktor Hamburger reporting on the effects of limb extirpation in chick embryos. My project had barely started when Giuseppe Levi, who had escaped from Belgium invaded by Nazis, returned to Turin and joined me, thus becoming, to my great pride, my first and only assistant. The heavy bombing of Turin by Anglo-American air forces in 1941 made it imperative to abandon Turin and move to a country cottage where I rebuilt my mini-laboratory and resumed my experiments.

—Rita Levi-Montalcini, Nobel Lecture, 1986

OVERVIEW

In this lecture, we'll learn

- ▻ The basic structure of the nervous system
- ▻ The principles of neural development
- ▻ The role of neuronal activity in critical periods for brain plasticity

The Nervous System

What we normally call the brain is actually part of the nervous system, which is pretty much a hollow tube of tissue that has been stretched and inflated at its top end. This tube contains cerebrospinal fluid, which brings oxygen and nutrients to all our neurons and glia and removes their waste. The nervous system is divided into the central nervous system (CNS) and the peripheral nervous system (PNS) (box 2.1). The CNS is itself subdivided into the brain and spinal cord, connected and protected by our skull and spine (figure 2.1). Since the brain is the largest part of the nervous system in mammals, we often use the word "brain" to refer to the entire nervous system, but it is technically not accurate, although we all do it.

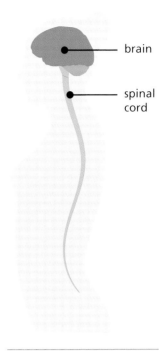

2.1 The **central nervous system (CNS)** is composed of the brain and the spinal cord.

The brain itself is subdivided into regions, grouped, from top to bottom, into forebrain, midbrain, and hindbrain, which then leads to the spinal cord (figure 2.2). The forebrain has two cerebral hemispheres that, in humans, balloon into an enormous cerebral cortex, which has a large (about a square meter) thin (about two millimeters thick) surface folded into winding turns (called *circumvolutions*) with grooves (*sulci*) and ridges (*gyri*), so it all fits into the skull (figure 2.3).

What does the cortex do? It is likely a universal computer, a biological Turing machine that, through learning, can adapt to solve any problem. Its enormous size in humans, relative to other animals, may explain our flexible behavior and associated evolutionary success. Beneath the cortex are the basal ganglia, involved in learned movements; the hippocampus, necessary for memory storage and navigation; and the amygdala, a hub for emotions.

BOX 2.1: Subdivisions of the mammalian nervous system. Conserved structures with specific functions form the nervous system in all mammals

Central nervous system (CNS):	Peripheral nervous system (PNS):
Brain	
Forebrain	Dorsal root ganglia (DRG)
Cerebral hemispheres	Autonomic nervous system
Cortex	Sympathetic division
Basal ganglia	Parasympathetic division
Hippocampus	Enteric nervous system
Amygdala	
Diencephalon	
Thalamus	
Epithalamus	
Hypothalamus	
Midbrain	
Hindbrain	
Medulla	
Pons	
Cerebellum	
Spinal cord	

At the center of the forebrain you have the diencephalon, a series of important nuclei that include the thalamus, which is involved in sensory processing and attention; the epithalamus, regulating our circadian rhythms; and the hypothalamus, a central station for emotions and hormones.

Below the forebrain lies the midbrain, a small but important series of nuclei controlling eye movement, reflexes, and sensory information from the head. Below the midbrain we find the hindbrain, another critical part of the brain, which includes the medulla, containing nuclei that control breathing, heart rate, and digestion; the pons, involved in control

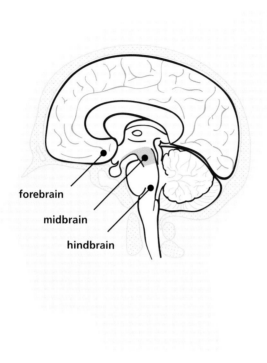

2.2 The **brain** is divided into forebrain, midbrain, and hindbrain.

of movements; and the cerebellum, a large structure with an enormous number of cells that are key for equilibrium, learning skilled movement, and motor memories. Moving downward from the hindbrain, we reach the spinal cord, a long tube of neural tissue that communicates and controls the rest of the body through sensory and motor circuits, giving rise to reflexes and motor actions.

The CNS also has two important structures serving as outposts. One is the retina, located at the backs of our eyes, processing all visual information before it is sent to the brain. The other is the olfactory bulb, at the top of our nasal cavity, which processes smells.

In addition to the CNS, our nervous system contains the peripheral nervous system (PNS), which is located outside the tube and structured

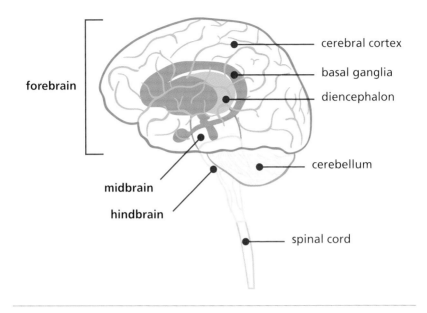

2.3 Major **anatomical subdivisions of the CNS**.

into a series of ganglia dispersed throughout the body. These include the dorsal root ganglia (DRGs; also known as the *somatic* PNS), which bring sensory information from the skin into the spinal cord; the autonomic nervous system, with its sympathetic and parasympathetic divisions, which serve as motor output for many of our emotional responses; and the enteric nervous system, a relatively independent part of our nervous system that controls our digestive tract, apparently independently of our input.

This basic design of anatomical subdivisions, each with a particular function, is conserved across evolution. All mammals have essentially an identical nervous system with all these same parts, which are larger or smaller depending on the species. The fact that the structure is similar across species implies great similarities between our brain function and that of other mammals, likely also including a similarity in basic emotional and computational properties.

Nowhere is the universality of evolutionary principles in neurobiology clearer than when we look at how the brain develops. The principles of how the nervous system is laid out are conserved across evolution, with some mechanisms even being the same all the way from cnidarians to humans.

How the Nervous System Develops

Remarkably, the brain actually assembles itself. How can this sophisticated and complicated system put itself together without anything instructing or directing it? Although many mysteries remain about exactly how this happens, the basic idea is that the nervous system develops through a stereotypical set of stages. It follows a step-by-step *temporal logic* of causation whereby what happens at a given time determines what happens next, like a beautifully choreographed ballet that unfolds in space and time. Thus, the apparently insurmountable task of building a brain is broken down into manageable steps—the stages of neural development—each of which is turned on by the previous step (box 2.2). It is all about timing, things happening in the right place at the right time.

BOX 2.2: **Stages of neural development.** Unfolding in time and space, the earlier stages are mostly genetically determined, whereas the later ones depend more on the outside world.

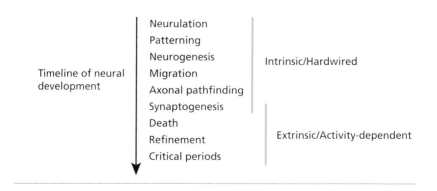

Neurulation and Patterning

The first step in brain development is the formation of the neural tube. This process is called *neurulation*. The CNS starts as a tubular infolding from the ectoderm, the outermost layer of the embryo. Our neurons thus originate from the same epithelium that makes our skin. The neural tube then invaginates, pinches off, and floats inside the embryo (figure 2.4 left). Essentially all our neurons are born in the walls of this tube, and their proliferation makes the tube grow, elongate, and change shape. The front becomes larger, ballooning into the future forebrain (figure 2.4 right), and you can start to recognize different parts of the future nervous system neatly divided into three bulges that become the forebrain (front), midbrain (middle) and hindbrain (back).

Now comes the *patterning* step, when a uniform neural tube becomes specialized into regions. Neurons in different parts of the CNS become

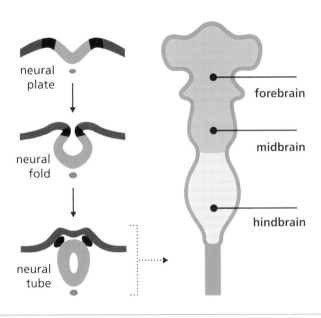

2.4 Neurulation occurs by invagination of the ectoderm, resulting in the formation of the neural tube, which takes different shapes along its axis (right).

differentiated from one another, acquiring a specific molecular identity depending on where they're located. Incidentally, this is how each part of the body develops: by molecular patterning depending on the embryonic position.

Tissue is spatially patterned through key regions in the embryonic nervous system that serve as developmental "organizers." That means they secrete molecules that can trigger, or "induce," a specific change in neighboring cells. Developmental induction was discovered by Ethel Browne, a graduate student at Columbia University in what was to become my own department. In 1909, Browne showed that you could induce a second body axis in a hydra by transplanting a specific region of a donor animal into a host. In later work, and without acknowledging her, Hans Spemann and Hilde Mangold repeated her experiments in the frog embryo, demonstrating that induction also occurs in vertebrate development, an experiment that earned Spemann a maybe undeserved Nobel Prize.

The way organizers work was explained in 1966 by another pioneering woman scientist, Hildegard Stumpf, who proposed that graded distribution of inductive molecules within tissues accounts for developmental patterning (figure 2.5). By diffusing in three dimensions through the embryo, these "inducing" molecules create molecular gradients, which serve as a kind of three-dimensional chemical GPS system. Imagine you are a cell in the embryo. You know precisely where you are because you can read these molecular gradients by measuring the concentration of each of the molecules that define three Cartesian coordinates. These molecular coordinates are unique to you; your neighbors experience a slightly different concentration of these inducers, thanks to the laws of diffusion, as they sit either upstream or downstream of the gradients. So any cell located in the embryonic nervous system receives different concentrations of specific inducing molecules, which create spatial gradients. Since Stumpf's findings, organizers have been found everywhere, orchestrating embryonic development in all of nature.

By creating the gradients of inducing molecules and taking advantage of their diffusion, nature is using the physics of the system to encode spatial information. Not only is it absolutely elegant that nature uses

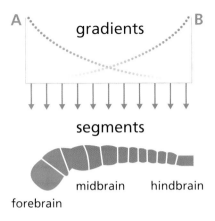

2.5 Diffusional **molecular gradients** provide positional information to the neural tube, locally turning on transcription factors (arrows) that can pattern the neural tube and generate cell types.

diffusion to pattern space, but, even more fascinating, these molecular gradients often follow Cartesian coordinates! Since Euclid, we have defined space as falling into three orthogonal axes: x, y, and z. We use these so-called Cartesian axes mathematically to determine any position in three dimensions with three numbers, describing the corresponding position of any given point along the axes.

The embryonic nervous system does the same thing but with molecules. Molecular gradients define the three Cartesian axes with respect to the body of the embryo: rostro-caudal (from head to tail), dorsal-ventral (from back to front), and medio-lateral (from midcenter to side). Evolution uses Euclidean geometry! But instead of numbers, cells count molecules—actually, concentrations of molecules. Moreover, these molecular gradients are conserved in evolution; the same molecule specifies rostro-caudal gradients all the way from cnidarians to humans. And who sets these gradients up in the first place? They are laid out in the zygote itself from the fertilized egg. And how does the fertilized egg know? The fertilizing sperm starts it all, as it sets the polarity of the embryo, which is defined with respect of its point of penetration. That single small event sets up a molecular machinery that patterns and creates the body in three dimensions.

But how do you turn a molecular gradient into a differentiation signal? Cells—neurons included—translate these external molecular signals into intracellular cascades that activate transcription factors. These transcription factors are DNA-binding molecules that regulate gene expression, turning blocks of genes on or off. Using transcription factors, or combinations of them, cells acquire a specific fate, which you can now measure quantitatively using single-cell transcriptomics as a molecular fingerprint that is specific for each cell type. So you build cell types with a transcription factor code.

We have about a thousand cell types in our nervous system (over a hundred in the mouse visual cortex alone), and it is likely that each of these cell types becomes so by expressing a specific combination of transcription factors during development. This transcription factor code is starting to be deciphered. Again, this process is not unique to the nervous system, as transcription factors are also used to build subtypes of cells throughout the body, demonstrating the universality of nature once more.

Neurogenesis and Migration

Now that we have neatly molecularly partitioned our brain into regions and segments and use this molecular code to neatly build neuronal (and glial) cell types, we come to the next step, *neurogenesis*, in which we generate billions of new neurons (figure 2.6). This process occurs mostly in the inside epithelium of the neural tube, the so-called ventricular zone. There, near the cerebrospinal fluid, as their primordial nourishing media, neuronal progenitor cells divide like clockwork. They move up and down, and every time this occurs, they split and generate another neuron. It's like a machine that generates neurons with every turn of the crank. That's how glia are generated as well, also in vast numbers.

These first neurons get pushed away by newer-born neurons from the center of the tube toward its periphery, gradually increasing the thickness of the neural tube. That's how the nervous system turns from a thin

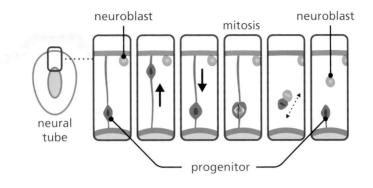

neuroblast

mitosis

neuroblast

neural tube

progenitor

2.6 Neurogenesis occurs in the ventricular zone of the neural tube and involves oscillatory movements of progenitor cells leading to a cell division, and generating a new neuroblast with each cycle.

tube into a thick tissue. Besides being pushed away, neurons also migrate. Most neurons migrate by climbing up radial glia fibers, which extend across the developing neural tube from the ventricular zone to the top (figure 2.7). This massive *migration* step is fascinating. Why aren't neurons born where they need to be in the first place? Why do neurons need

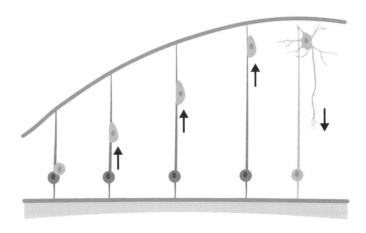

2.7 New neurons (red) **migrate** along radial glia (gray) until they reach their destination, stop, and generate an axon (right).

to be born precisely next to the ventricle, in the inner epithelium of the neural tube, instead of anywhere else in the nervous system? There must be something critically important about the ventricular zone, perhaps having to do with the primary cilia of the progenitor cells that protrude into the central cavity. In any case, after patterning, neurogenesis, and migration, we finally have our full complement of neurons, positioned in their correct places ready to connect.

A final comment: neurogenesis in humans is a prolonged process, which explains not only our much larger brains but also our protracted period of neuronal development. We are very slow to develop compared with other animals. This delaying or slowing of physiological, or somatic, development is called *neoteny*, and the slow construction of our brains during development, as we will see later, could help us achieve greater computational and cognitive functioning.

Wiring and Pathfinding

Now comes perhaps the most critical step in neural development: *axonal pathfinding*. In this remarkable feat, developing neurons extend an axon that often leaves the area where the neuron has settled and navigates within the embryonic nervous system, searching for greener pastures. The axon somehow finds its way through the tissue until it reaches a target neuron and connects with it (figure 2.8). Unlike the development of other organs, which also proceeds through similar steps of patterning and cell division, axonal pathfinding happens only in the nervous system. In fact, you could argue that it is the wiring that makes the nervous system special. The brain is all about wiring.

Axonal pathfinding is done by a specialized structure at the tip of the growing axons, the *growth cone*. This fascinating hand-like structure was discovered by Cajal (see the drawing that opens this lecture), who correctly imagined that it serves for navigation by following chemical signals. Growth cones are amazing navigators. For example, some neurons from our motor cortex send an axon to the lower part of the spinal cord to connect with neurons located there that control our leg muscles. These

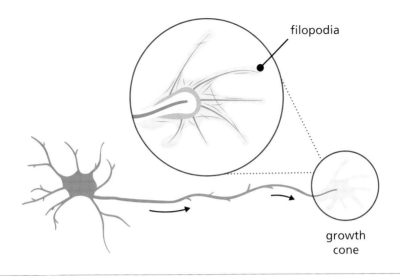

filopodia

growth
cone

2.8 Axonal pathfinding is done by growth cones navigating the developing CNS following chemical cues

connections are extremely precise; an axon connects with a specific neuron that moves one muscle, but not with its neighboring neuron a few microns away that moves a different muscle.

Now let's put this in number form. Let's say that a typical neuron in our motor cortex has a cell body of about 30 μm in diameter. To reach the bottom of the spinal cord, its axon is typically longer than 1 m. Now assume you are that neuron. If your body were the size of that neuron, this axonal length would be similar to stretching your arm to send your hand on a journey as long as 70 km to find one specific person and shake hands with them—and only with that person and not their neighbor. This is remarkably specific navigation.

How would you extend your arm for 70 km—let's say from one city to the next—and find the right neighborhood, the right street, and the right apartment and end up shaking hands with the right person and not with their roommate? The way we do this, the same way we navigate as humans, is by breaking the journey into steps. To get from our home to our friend's home in the next city, we first get out of our building, leave our neighborhood, and travel out of our city. We then reach intermediate milestones along the way and follow them until we reach our destination. The GPS in your phone works just like that: it provides you step-by-step instructions. The growth cone navigates the same way: breaking the journey into many steps and reading instructions from one milestone to the next until reaching its destination.

How does the growth cone find and recognize each new milestone? Again, the answer is a molecular trick. Remember the molecular GPS that defines the position in three dimensions of every part of the embryonic nervous system? Because they have distinct molecular identities, neurons in different locations also have different molecular markers on their surfaces; in other words, they secrete molecules that can be read and recognized by the axon's growth cone humming around. It's like a molecular zip code map that the growth cone can read. This is known as the *chemoaffinity* hypothesis: the "fingers" of the growth cone follow a trail of molecules laid out in the tissue and can "smell" them to determine where they are heading. The growth cone is dumb; it's just a machine that can move forward or backward, left or right. A bit like when you are following your car's step-by-step GPS navigational instructions and stop thinking about how to get where you're going.

A growth cone moves forward if it encounters an attractive molecular cue, a chemo-attractant, and backward if it finds a chemo-repellant. The combination of attractive and repulsive chemical cues eventually directs the growth cone to its correct target. Growth cone navigation looks a little messy when you watch it on video, as if it were a hand moving its fingers around, looking for something, but it works reliably and robustly. A lot of stops and starts, turning left and right, but eventually the growth cone gets precisely where it's going.

Synaptogenesis

Now comes yet another beautiful step: *synaptogenesis*, the formation of synaptic connections. Why beautiful? It has nanometer precision, and is also self-assembled. There is no one providing directions.

A synapse is formed when a growth cone finds the body or dendrites of its target neuron. Together, they start an intricate dance, exchanging molecular signals back and forth (figure 2.9). This process has been best studied in the synaptic connection between the axon of a motor neuron and a muscle cell, which is called the *neuromuscular junction*. Because of the ease of experiments examining this system, it has served as an excellent entry point into understanding synapses.

Synaptogenic signals going back and forth between the pre- and postsynaptic cells are of many types. Some molecules can trigger (induce) specific changes in the recipient cell. Inducing molecules from the axon generate expression and clustering of neurotransmitter receptors, building a sophisticated postsynaptic structure that is exactly opposite to the axon. On the axon side, the growth cone turns into a mature presynaptic terminal, and its hand shape becomes something like a fist. Presynaptic terminals, as we will see in the next lecture, are also incredibly sophisticated nanomachines. Some of the signaling that occurs between the pre- and postsynaptic sides is likely also mechanical. Like good dancers, they

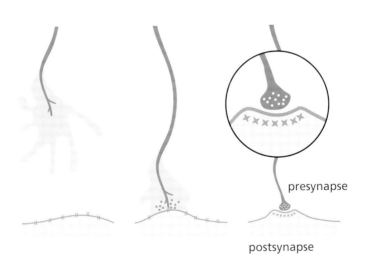

presynapse

postsynapse

2.9 Synaptogenesis involves intricate communication between the growth cone and the target neuron.

feel each other and react to each other's movements. Once synapses form, they are pretty stable, although there is still room for later modifications. These subsequent changes, called *synaptic plasticity*, occur during the rest of the life of the animal and probably are one of the key mechanisms for learning.

Neuronal Death

After the dancing of the growth cones and the happy marriage of the pre- and postsynaptic structures comes the tragedy. Let's talk about death. Actually, not death, but suicide, one of the strangest steps in neural development. The story starts with the painstaking work of Viktor Hamburger, Rita Levi-Montalcini's mentor. A student of the controversial Spemann, Hamburger was a German developmental neurobiologist interested in the development of the neuromuscular junction. After being expelled from the University of Freiburg for being Jewish, he moved to St. Louis, where he started to carefully count the numbers of neurons in spinal cords of chick embryos at different stages of development. And what he found was very strange. According to his counts, there were about 50 percent fewer motoneurons at the end of development than at earlier stages (figure 2.10). Hamburger argued that this was caused by massive cell death of neurons, and he postulated that nature used this culling technique to match the number of spinal cord motoneurons to the number of muscles.

earlier   later

2.10 Massive **neuronal cell death** occurs in the spinal cord during development. Green dots are motor neurons.

The debate as to why that die-off happens is still unresolved, but we learned a lot about how it happens, as is often the case in biology, where you understand the phenomenon and mechanisms before you understand the reason behind it. Hamburger postulated that as connections are formed, some of the signals between the postsynaptic cell and the presynaptic axon take the form of molecules that support the cell. These trophic factors, or *neurotrophins*, are necessary for the survival of the presynaptic neuron. The first neurotrophin was discovered in an amazing set of experiments involving persistence but also sheer luck and serendipity carried out by two disciples of Hamburger, Levi-Montalcini and Stanley Hoffman. Working in Hamburger's department, and with characteristic persistence demonstrated by the opening quote describing how Levi-Montalcini secretly continued experiments during the war, they purified the first neurotrophin, which they called the "nerve growth factor," or NGF. To purify NGF, they needed a reliable source of proteases and chose snake venom, which is full of them. In an amazing turn of events, snake venom itself was found to be a concentrated source of NGF! The rest was downhill from there: NGF promoted survival of presynaptic neurons in sympathetic ganglia, so, consistent with Hamburger's hypothesis, it could serve to match the number of presynaptic neuron and postsynaptic targets. The more postsynaptic cells, the more NGF is secreted and the more presynaptic neurons survive.

NGF was the first of many neurotrophins necessary for the survival of neurons in the peripheral and central nervous system. Indeed, although we still don't have all the numbers, a massive die-off seems to occur throughout the course of brain development. It is likely that the majority of the neurons in our brain die during development.

How do all these neurons die? The way neurotrophins promote neuronal survival appears to be by blocking the apoptosis pathway. Apoptosis is a form of cellular suicide, or programmed cell death. The bottom line is that developing neurons are designed to die by suicide unless they are rescued by a neurotrophin. Talk about the tragic fate of being a neuron! It is remarkable to think that the way the nervous system is put together is sort of backward, by generating extra numbers of neurons, which then get culled as development proceeds. Just like sculpting an object out of a block of marble.

Synaptic Refinement

Sculpting does not end with cell death. In the next step in development, the synaptic connections from the neurons that survive the culling are also heavily pruned. Again, this pruning was also first found in the neuromuscular junction. In 1907, Francisco Tello, a disciple of Cajal, cleverly noticed that early in development, each muscle fiber was innervated by many axons, yet in the adult, each muscle fiber was innervated by precisely only one single axon (figure 2.11)—not two or three or zero, but always one. He argued that connections were being lost, and that this process of elimination occurred through competition among axons for muscle fibers. For every muscle fiber, there is a winning axon and several losers. Just like in the game of musical chairs, when connecting neurons to muscles, you have more dancers than chairs.

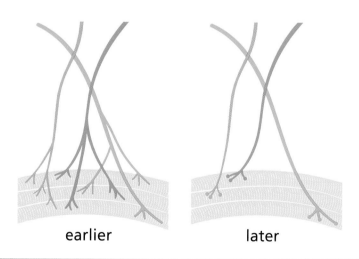

earlier   later

2.11 Pruning of extra connections occurs during muscle development, resulting in each muscle fiber being innervated by only one axon.

How important is this process of elimination? It's widespread not only in neuromuscular junctions but in the brain itself. Current estimates are that at least 50 percent of the synapses in the developing central nervous system disappear after birth. This enormous synaptic elimination is partly controlled by neuronal activity. Neurons that act jointly appear to maintain the synapses between them. Meanwhile, neurons that are not coactive and fire independently, not in concert with other neurons, lose their synapses. In other words, use it or lose it.

This principle—which, as we learned in the last lecture, was formalized by Donald Hebb in 1946 to explain how neural circuits store memories—also applies to the developing nervous system and could be behind the massive pruning of connectivity. So at the end, the developing animal is actually sculpting its brain. Neuronal activity, driven by early experiences, is the sculptor.

Neuronal Activity and Critical Periods

Activity-dependent refinement happens at particular times during development. This brings us to a seminal experiment that was carried out by Torsten Wiesel and David Hubel in 1963. These experiments are close to my heart because I got my PhD with Wiesel many years later, so he is my "scientific father." When recording the activity of neurons in the visual cortex of young kittens, they discovered that a brief period of visual deprivation during development forever altered neural circuits in the visual cortex. If they sutured shut one of a kitten's eyelids and let the animal develop, the closed eye lost its innervations to the cortex and the animal became blind in that eye. This blindness was due to an imbalance in the activity of both eyes, which compete for the same cortical territory, just like motoneuron axons compete for the same neuromuscular junction. In the visual cortex, this competition normally leads to axons sorting themselves out, with each eye controlling one patch (that is, a "column") of cortex (figure 2.12). But more interestingly, if the same eyelid suture (monocular deprivation) was carried out in an adult cat, nothing happened. So the effect of monocular deprivation happens only

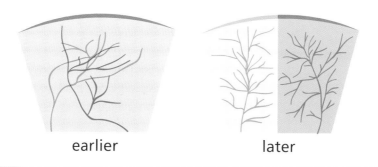

earlier   later

2.12 In visual cortex, refinement of connections occurs during a **critical period** of postnatal development, in which axons from both eyes compete for the same cortical territory.

during early development, and once the competition period has passed, the cortex is crystalized.

These classic experiments demonstrated how early experience can craft brain connections, defining a "critical period" of neural development during which neuronal activity instructs which synapses make it and which are lost.

Developmental critical periods are not unique to the visual system, as later work showed that they happen almost everywhere in the brain. In fact, most brain functions are shaped during critical periods, which, depending on the part of the brain, occur at different times during development. For example, in addition to the ocular dominance critical period necessary for binocular vision, there are critical periods for walking, learning a language, acquiring musical pitch and skilled movements, among other functions.

We are not alone in this process, as many other animals also experience critical periods. Remember that famous picture of the ducklings that were imprinted during hatching on the investigator Konrad Lorenz and which followed him as if he were their mother? The reason is that as soon as they hatch, ducklings experience a critical period for identifying mom, causing them to imprint on their mother and follow her faithfully until they become adults. It's preprogrammed behavior.

Social interactions also have critical periods, as demonstrated in experiments with monkeys that were raised in isolation, without their mothers or other monkeys. Those animals were asocial later in life. In fact, the existence of critical periods in humans demonstrates the importance of early education and exposure of children to enriched environments. In this tension of nature (that is, genetics plan) versus nurture (activity-dependent sculpting) that shapes our nervous system in development, we should help as much as we can to ensure that environmental conditions do not become limiting. As the Wiesel and Hubel experiments showed, if you miss the critical period, your brain becomes hardwired, and there is no way back.

The mechanisms that regulate critical periods during development are starting to be understood. Neurotrophins appear to play a major role; not only do they help neurons survive, but they also protect synapses from being pruned away. Another key mechanism seems to be the N-methyl-D-aspartate (NMDA) glutamate receptor, a molecular sensor that can detect the coincidence in time of the input and output of a neuron. That is, it switches on only when the neuron receives synaptic inputs and is firing at the same time. When both things happen at the same time, the NMDA receptor (NMDAR) opens and lets a significant amount of calcium into the cell, turning on intracellular pathways and gene expression. This can help explain why the synapses from the winning axon are protected (because their strength enables them to open NMDARs) but those from the weaker axon are eliminated (as they are too weak to open NMDARs).

Both mechanisms, neurotrophins and NMDARs, are also used in the adult nervous system to regulate synaptic plasticity. The only difference between the plasticity that goes on during critical periods and in adulthood is one of magnitude; plasticity remains but is less prominent, as someone like me who has tried to learn a foreign language in adulthood, past the critical period for language, can attest.

So for mysterious reasons, evolution has made developing brains extremely plastic during these critical periods as compared with adulthood. Why is that? Pruning of neural circuits in an activity-dependent manner during development sort of makes sense if you think about it. The nervous system first wires itself up by making a lot of extra connections (in fact, even a lot of extra neurons) and then prunes away unnecessary ones by using the patterns of activity that are generated in response to the environment in which the young animal lives. This is just like the FPGAs that engineers use and we discussed in the previous lecture. Neural activity acts as the pruning shears. Using neural activity, the circuit can fine-tune its functional properties: from sharpening topographic maps, to responding faster and more precisely to sensory stimuli, to adjusting the nervous system to the rest of the body—everything can be improved by letting go of the extra "fat" of unnecessary cells and connections. By pruning connections through environmental influences, nature guarantees that the nervous system is precisely matched to the environment in which the animal lives. Thus, your language neural circuits are matched to the language that is spoken in the place you are born. This happens with many of the behaviors that we develop and nicely explains why brains are so plastic during development.

This sounds like a good thing, but why shouldn't brains continue to be plastic during adulthood? Wouldn't that be great, making us smarter and more flexible in our behavior and enabling us to continuously learn all kinds of things or languages effortlessly? What's wrong with being able to learn languages, for example, one after the other, with the same ease as when we learned our mother tongue? This is a deep mystery. No one knows why the critical periods need to close. Cajal used to say that during development, neural circuits are like mortar that sets for good. This metaphor may be right on since the laying out of axons' insulation (myelin, which we will learn about soon) and extracellular molecules that encase neuronal cell bodies (perineuronal nets) could be mechanisms that shut off the critical period. But why—not how—does it close? One possibility is that maintaining a high level of plasticity may interfere with the daily function of neural circuits in adulthood. Perhaps it's not a good idea to be able to learn too much, because it could destabilize what we already know. This could create disturbances that could be maladaptive. What types of disturbances? Let's look at humans.

As a species, we are notoriously adept at learning even after adulthood, probably the champions in the animal world in this ability. But we are also unique among animals in suffering mental diseases, which are present in a significant proportion of the population and which you would think should have been eliminated by evolution, because they are maladaptive.

Perhaps the fact that humans suffer mental diseases is the price that we pay as a species for having particularly plastic brains.

▼ RECAP

Conserved structures with specific functions form the nervous system in all mammals, and these structural modules are built during development via a progression of steps. Using molecular mechanisms, these modules, often organized into maps, self-assemble in stages, of which the initial ones are intrinsic and genetically determined, whereas the latter ones are regulated by neuronal activity.

The CNS starts as a tubular infolding from the ectoderm that pinches off and elongates. Parts of it become differentiated from each other, acquiring a specific molecular identity depending on where they're located. After the neural tube is formed and patterned, neurons and glia are born and generated in great numbers. After that, neurons migrate to their final positions and start extending axons to connect to other neurons. After these initial connections are made, many neurons—perhaps most—commit suicide if they lack successful contact.

Following this massive neuronal die-off wave, another wave of massive elimination of connections occurs. This refinement of the wiring depends to a large extent on the patterns of neuronal activity in the embryo and the young animal during a critical period of neuronal development. In this way, our nervous system is carved to match the environment in which we grow up, so nature does not need to specify absolutely everything as it builds our brains. After that, our neural wiring stabilizes, and we are ready to live in the world. Our brain changes very little for the rest of our lives. We still have some plastic changes in connectivity, but to a much lesser extent than when connectivity occurs during those critical periods of development. In fact, as adults, our CNS has very limited regeneration ability. With some exceptions—in parts of the olfactory system and the hippocampus—the neurons that you are born with are all you get. If you lose them, you will not get them back.

Thus, neural development proceeds from being genetically controlled to being dominated by neuronal activity. In a way, the initial stages are preprogrammed via both genetic and epigenetic instructions. The neurons just follow their playbook and become what their progenitors instructed them to become. It's a fixed, *intrinsic* developmental plan. With his characteristic wit, Sydney Brenner called this the "European plan": what you become in life is determined by your parents. Meanwhile, in later stages, the development of neurons, and even their survival, depends critically on how the neurons are connected, whether or not they fire in concert with other neurons and whether those ensembles are successfully activated from the outside. Brenner called this *extrinsic* plan the "American plan": what you become depends on your connections and activity; that is, your surroundings and your interactions with them. Importantly for our cognition, later stages are particularly protracted in humans; they can last for more than a decade. That is extremely unusual, even for primates. Why do we educate our children for so long? Because their brains are still being sculpted and refined, chiseled away by the extrinsic influences (i.e., teachers, parents, peers, and the media).

Now that we understand how the system is put together, how does it work? In the next lecture, we'll get into the nitty-gritty: what type of cells are the neurons, how do they communicate with one another, and how do they build neural circuits?

Further Reading

Brenner, S. 2001. *My Life in Science*. BioMed Central. A personal story of one of the most influential scientists in the twentieth century and the person that changed my life.

Browne, E. 1909. "The Production of New Hydranths in Hydra by the Insertion of Small Grafts." *Journal of Experimental Zoology* 7: 1–37. A seminal experiment by a brave graduate student who was not recognized in her time, which showed for the first time the induction by a transplant of a secondary axis of polarity in a host embryo.

Purves, D., and J. W. Lichtman. 1985. *Principles of Neural Development*. Sunderland, MA: Sinauer Associates. A personal favorite, a lucid account of the phenomenology of neural development.

Sunkin, S. M., L. Ng, C. Lau, T. Dolbeare, T. L. Gilbert, C. L. Thompson, M. Hawrylycz, and C. Dang. 2013. "Allen Brain Atlas: An Integrated Spatio-Temporal Portal for Exploring the Central Nervous System." *Nucleic Acids Research* 41: D996–D1008. A molecular map of the brain, gene by gene, brain region by brain region.

Stumpf, H. 1966. "Mechanism by Which Cells Estimate Their Location Within the Body." *Nature* 212: 430–31. Another seminal paper, also from another woman who was not recognized, that explains how molecular gradients serve as GPS coordinates for cells in the embryo.

Swanson, L.W. 2002. *Brain Architecture*. Oxford: Oxford University Press. A scholarly description of the anatomical basis of the nervous system and its developmental and evolutionary logic.

Tabula Sapiens Consortium. 2022. "The Tabula Sapiens: A Multiple-Organ, Single-Cell Transcriptomic Atlas of Humans." *Science* 376, no. 6594. https://www.science.org/doi/10.1126/science.abl4896. A landmark study that uses single-cell trasncriptomics to map cell types across human tissues and organs. A new era lies ahead in which we will identify all the subtypes of cells in the body (and the brain) and decipher their transcription factor code. This could enable selective manipulations of cell types for therapy and synthetic biology.

Wiesel, T. N., and D. H. Hubel. 1963. "Single-Cell Responses in Striate Cortex of Kittens Deprived of Vision in One Eye." *Journal of Neurophysiology* 26: 1003–17. A simple yet hugely influential paper describing how the environment during a critical period of development can profoundly shape brain function.

Yuste, R., et al. 2020. "A Community-Based Transcriptomics Classification and Nomenclature of Neocortical Cell Types." *Nature Neuroscience* 23, no. 12: 1456–68. https://www.nature.com/articles/s41593-020-0685-8. A grassroots proposal to systematically classify and name all the neurons in the cortex.

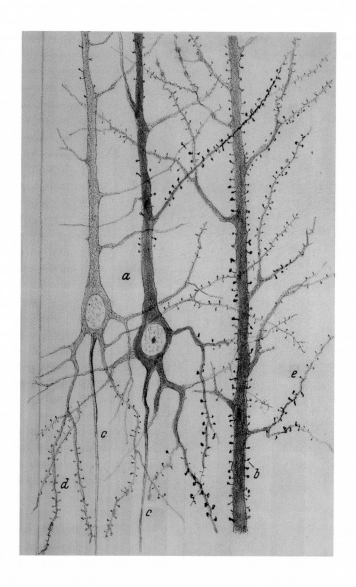

Pyramidal cells from the cerebral cortex of a rabbit. Pyramidal neurons are covered with dendritic spines, small protrusions that are likely building blocks of our thoughts and memories. Courtesy of the Cajal Institute, Cajal Legacy, Spanish National Research Council (CSIC), Madrid, Spain

LECTURE 3: NEURONS

Like the entomologist in search of colorful butterflies, my attention has chased, in the gardens of the grey matter, cells with delicate and elegant shapes, the mysterious butterflies of the soul, whose beating of wings may one day reveal to us the secrets of the mind.

—Santiago Ramón y Cajal, *Recollections of My Life* (1917)

OVERVIEW

In this lecture, we'll learn

▶ How neurons work
▶ Types of neurons
▶ Principles of cellular neuroscience

Neuroscience is over a hundred years old. It grew out of anatomy and physiology. Armed with anatomical methods like the Golgi stain, which labels neurons in all their glory, or physiological methods like the micro-electrode, which records the activity of individual neurons, generations of researchers plowed ahead, taking apart the nervous systems one neuron at a time and studying each of them intensely. Those studies have led to a century of progress in cellular neuroscience, aided by a molecular revolution in the last decades as methods of molecular biology and genetics have been systematically applied to disentangle the neuronal molecular machinery. In this lecture, we will summarize some of those findings. It's hard to fit a century of progress into one lecture, so for more detailed coverage, I recommend you read the excellent textbooks mentioned in lecture 1, which cover cellular and molecular neuroscience in encyclopedic detail.

Neurons

The nervous system is built with two types of cells: neurons and glia. All neurons are essentially built the same: they have dendrites, where synaptic inputs arrive; a soma, where those inputs are summed; and an axon, which tallies that integration and generates the action potential, or spike, an electrical signal that propagates to the next neuron through a synapse (figure 3.1). That process is repeated over and over.

The nervous system is a gigantic machine filled with these input–output elements. Like bucket brigades, strings of these neurons form neural circuits, which are used to build reflexes, and all the magic we were talking about in the first lecture. To understand this in more detail, let's take a neuron apart and march our way through it step by step. We start at the axon and axonal terminal, then move to the synapse, then we will cross to the dendrites of the postsynaptic cell and review dendritic integration, to arrive back at the soma and the generation of the action potential. In this journey, we will also ponder the biological and computational purpose of the mechanisms that neurons use in their daily function. We will try to understand not just how neurons work, but why they work the way they do.

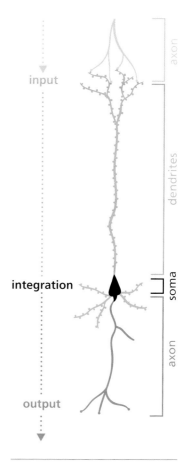

3.1 Neurons receive inputs through their dendrites, integrate them electrically in their somata, and send their outputs down their axons.

Ion Channels

What makes neurons special compared with other cells in the body? They are electrical devices. They have a semipermeable membrane, which lets in some ions but not others. Neurons are chock-full of potassium, with very little sodium or calcium, whereas the outside media has the opposite situation: a lot of sodium and calcium with little potassium. This ionic asymmetry builds a significant electrical potential across the plasma membrane, the ionic equilibrium potential, which is then cleverly harnessed and manipulated by nature for electrical signaling.

Ion channels, proteins on the membrane that specifically let in or out some ions but not others, do this signaling. Ion channels are astounding molecular machines with a design that has been conserved through evolution. A part of them is electrically charged and sits inside the membrane, so it responds to electric fields, moving up and down the membrane as they ride the membrane potential waves. And this movement opens or closes a tiny tunnel through the membrane through which ions flow (figure 3.2). As noted, each ion channel has been designed to be particularly permeable to a given ion and not others. This selectivity is achieved

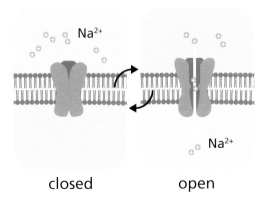

Na^{2+}

Na^{2+}

closed   open

3.2 **Ion channels** are membrane proteins that let ions flow into, or out of, the neuron, by opening a pore that is selectively permeable to a type of ion (sodium in this case)

by tunnels of the exact size for each type of ion and is formed by amino acids lying in the pore that form pockets that precisely match the dimensions of the ion. Ions flow one by one, in single file, at great speeds: a good ion channel can pass several thousand ions in a millisecond! And crucially, because ions are charged, these ionic fluxes not only respond to the membrane potential but also can change it.

The channels we have discussed so far are called the *electrically gated* ion channels and act quickly, responding within a millisecond of the membrane potential change. But there are also others that are instead turned on by biochemical signals inside the cell. Those are *second-messenger gated* ion channels, and they can be used for slower modifications of the membrane potential.

Action Potentials

At rest, neurons have a hyperpolarized potential, meaning that the inside of the neuron is about 65 mV negative to the outside, due to the ionic equilibrium potential just described. But this resting membrane potential can change quickly due to a combination of two types of ion channels, permeable for either sodium or potassium, which have opposite actions and together generate the action potential.

Action potentials are brief depolarizations of about 100 mV that last a few milliseconds (figure 3.3). They are generated when sodium channels open, letting positively charged sodium ions into the neuron. That depolarizes the membrane. This depolarization then trips potassium channels, which open, after a small delay, and let out positively charged potassium ions, which hyperpolarize the membrane back to where it was initially. Thus, the membrane experiences rapid depolarization followed by slower hyperpolarization, all in a few milliseconds. The shape of this action potential is stereotypical: it is nearly the same in all species. And, most importantly, the action potential propagates along the membrane, in this case the axon. How?

Imagine an action potential that happens in one position of the membrane of an axon. The depolarization trips open sodium channels in the

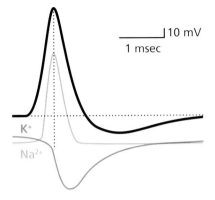

3.3 The **action potential** is a large and fast swing in membrane potential (black) generated by a depolarization caused by the opening of sodium channels (green), followed by a hyperpolarization caused by the delayed opening of potassium channels (red).

neighboring patch of membrane, which generate an action potential there. That trips the next membrane patch, and so on. But, after an action potential occurs, sodium channels inactivate for a brief period, called the refractory period. So once an action potential has passed through a patch of membrane, it cannot go back, since the sodium channels will be inactivated and will not open for a while. This ensures that the action potential always propagates forward, as a wave that propagates down the axon, but cannot return.

Action potential propagation happens at great speeds, up to several meters per second, enabling the rapid propagation of information in the nervous system. Since action potentials are all or none—meaning that they either happen or don't, and once they happen, they have the same size—this signal is digital. So this is digital electronics, and was invented by evolution more than a billion years before humans made their own version of it with transistors.

Not only did biology invent digital electronics, it also invented electrical cables. To enhance the propagation of action potentials in long axons, nature uses two tricks, just like we do with our own electrical cables. The first is to make axons thicker, which lowers the internal resistance, increasing the propagation of the current inside. The second trick is to wrap them in insulating material, increasing the membrane resistance, to block current from leaking out the membrane. This insulation is built by myelin, the processes from particular glial cells, the oligodendrocytes, that wrap themselves in layers around the axon. The combined effect of both tricks helps action potentials propagate farther without decrement. Evolution knows electrical engineering!

But why use action potentials at all? As all-or-none, digital signals are reliable and can propagate across great distances without loss of information. Because they're so fast and enhance their speed with myelin and large diameters, it's likely that their exact timing matters a lot. They could turn information into a temporal Morse-like code that is sent around the brain and deciphered by the recipient neurons. In other words, at its heart, the brain is a temporal machine.

Synapses

Neurons communicate with each other using synapses (figure 3.4). There are two types of synapses: electrical and chemical. Electrical synapses are made by gap junctions, which are special channels built between two neurons. They are formed by proteins on either side, called *connexons*, which lock together, making a tunnel between the cells. This enables ions to flow through so the action potentials or depolarizations of one neuron spread to the other. The function of electrical synapses is still poorly understood. They are used extensively during development, perhaps for pattern formation of tissues, coupling cells that will perform the same function. In the adult animal, they can synchronize the activity of neurons and enable neurons to share second messengers.

But the majority of synapses, at least in mammalian nervous systems, are chemical. In these synapses, the action potentials are transformed into a chemical signal, the neurotransmitter, which is then transformed back by the recipient neuron into a new electrical signal: the synaptic potential. Thus, these synapses carry out a transformation from electrical to chemical to electrical. This is quite peculiar. Let's take a closer look at how this works.

Axons end in presynaptic terminals. They are specialized structures a few microns in diameter that are filled with synaptic vesicles, storing

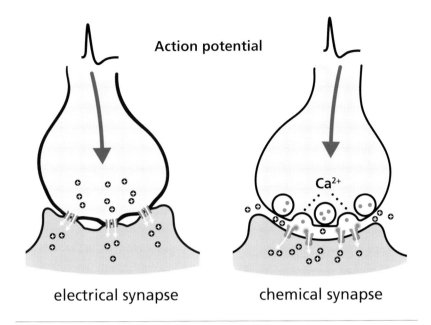

Action potential

Ca²⁺

electrical synapse   chemical synapse

3.4 Synapses serve to communicate electrical signals from one neuron to the next. They can be "electrical," by building a physical opening between the two neurons, so ionic currents flow between them, or "chemical," with the first neuron releasing a neurotransmitter that diffuses to the other side and activates receptors that generate an electrical signal in the second neuron.

extremely high concentrations of neurotransmitter. Synaptic vesicles become docked at the membrane, so they are ready to be released when an action potential invades the presynaptic terminal. When that happens, synaptic vesicles fuse with the plasma membrane of the presynaptic terminal and release the transmitter into the synaptic cleft. The cleft is an extremely narrow gap, about 20 nm thick, that separates the presynaptic terminal from the postsynaptic dendrite. The narrow cleft ensures that the transmitter diffuses very quickly and binds to receptors on the postsynaptic side.

There are many types of transmitters and their corresponding receptors. In the neuromuscular junction—the synapse between the axon of a motoneuron and a muscle fiber—the main transmitter is acetylcholine,

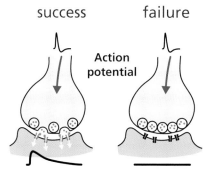

success   failure

Action
potential

3.5 Most brain synapses have **stochastic transmission**: sometimes an action potential releases the neurotransmitter and depolarizes the postsynaptic cell but sometimes it doesn't.

which binds to nicotinic acetylcholine receptors in the muscle. Neuro-muscular junctions are highly effective: an action potential always releases docked vesicles and always leads to muscle contraction. It's a fail-safe machine.

But synapses in the CNS are very different: they are stochastic. In other words, if an action potential reaches the presynaptic terminal, sometimes it releases neurotransmitters, but sometimes it doesn't (figure 3.5). The stochastic nature of chemical transmission in the brain is one of the big-gest mysteries in neuroscience. Why would you want to build a machine using elements that sometimes work and sometimes don't? This is not a small design problem; the rate of failure of synapses can be very high: a typical synapse in your cerebral cortex could fail more than 50 percent of the time. Fascinating! So why does the brain work as an essentially random machine, flipping a coin as to whether or not to transmit? That may also have to do with the way circuits work. As we will see in the com-ing lectures, if neural circuits work as neural networks, stochastic trans-mission could help them explore a potentially vast computational space effectively. Some random jiggling could allow neural circuits to avoid set-tling into activity patterns that may arise first in time, let's say in response to a stimulus, but not be an ideal match. So with stochastic transmission, the circuits could get out of this rut and find a better fit. Human metal-lurgists made a similar discovery long ago: they jiggled the temperature of the oven as the steel settled so it became harder and more flexible as the atoms found a different configuration, a process called tempering. Stochastic transmission could temper neural circuits. The neuromuscular junction, on the other hand, is exactly the opposite: it does not serve to compute but just acts. It receives the command to contract a muscle, previously computed by a network of neurons, and executes it faithfully, without messing around. That's why neuromuscular junctions are not stochastic and never fail.

And, finally, why do neurons have chemical synapses at all? Why not just use electrical synapses to pass around electrical signals? Gap junctions are indeed fast and effective, but you can make only so many before you electrically shunt out the neuron. Neurons are not infinite batteries, and every hole in the membrane diminishes their current. Meanwhile, with a single axon that ends in hundreds or thousands of synaptic terminals, there are no membrane leaks, and you can greatly amplify the message you want to send to the circuit. So one major advantage of chemical transmission is that the output of a neuron is greatly multiplied. That capability may have paved the way for the building of distributed circuits to implement emergent properties, since, ideally, you need to distribute the message to everyone. The design logic of these biophysical mechanisms of the neuron may ultimately rest at a higher, circuit level.

Receptors

CNS synapses mostly release glutamate, the main excitatory neurotransmitter in the brain. Glutamate binds to two types of receptors: *ionotropic* and *metabotropic*. Ionotropic receptors are fast-acting, and when glutamate binds to them, they open a channel that brings positively charged sodium into the cell (figure 3.6 top). Hence, they depolarize the cell, which is why they are considered excitatory, generating an *excitatory postsynaptic potential* (or EPSP). Ionotropic glutamate receptors can belong to one of three subtypes: N-methyl-D-aspartate (NMDA), α-amino-3-hydroxy-5-methyl-4-isoxazolepropionic acid (AMPA), or kainate, each with peculiar biophysical and molecular properties.

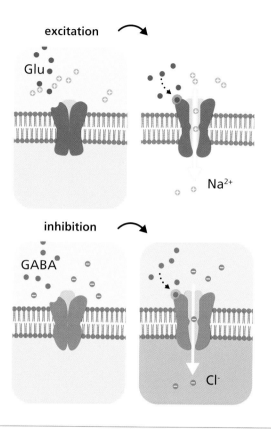

3.6 Receptors are membrane proteins that bind neurotransmitter and open to pass ions, leading to depolarization (glutamate, top) or hyperpolarization (GABA, bottom) of the postsynaptic cell.

Metabotropic receptors, on the other hand, are slow and coupled to second-messenger pathways. Thus, when glutamate binds to them, they trigger a biochemical cascade, which can secondarily open or close other ion channels. Depending on which channel they open they can depolarize or hyperpolarize the neuron. All glutamate receptors—in fact, all receptors—have a modular structure, built with subunits, and are conserved throughout evolution.

The second most common neurotransmitter in central synapses is GABA (gamma amino butyric acid), another amino acid that often has actions opposite to those of glutamate (figure 3.6, bottom). GABA receptors, also built with modular subunits and conserved in evolution, can be ionotropic and metabotropic, so their effects can also be direct or indirect. When GABA binds its ionotropic receptors on the postsynaptic side of the neuron, it often leads to an influx of chloride into the cell (figure 3.6, bottom). Now, because chloride is negatively charged, this hyperpolarizes the cell, which is why GABAergic synapses are considered inhibitory, generating an *inhibitory postsynaptic potential* (IPSP). But, depending on the stage of development and the exact position of the synapse, GABA can also have excitatory actions, because sometimes there is more chloride inside than outside the neuron. So when a chloride channel opens, chloride can leave the cell, making it more positive, depolarizing it. GABAergic receptors can also be metabotropic, leading to longer lasting effects.

Besides glutamate and GABA, many synaptic terminals release neuropeptides. These are highly diverse, belonging to a dozen different families and still poorly understood. They normally act through G-protein-coupled receptors, so they are metabotropic. In simpler nervous systems, like cnidarians and invertebrates, neuropeptides play a major role in behavior, specifying precise sets of movements, so one could argue that they are chemically encoding computations. The role of peptides in mammalian nervous systems is often underappreciated. In mammals, we know that neuropeptides are everywhere, often co-released with GABA. In some cases, they also control specific behaviors—reproduction and food intake, for example—but that is just scratching the surface; they likely play a major role in many behaviors.

Is there any logic behind all these transmitters? Neurotransmitters, neuropeptides and, more generally, molecules in neurons could implement computational variables. All these different transmitters, receptors and their subunits, ion channels, other proteins, second messengers, and so on are likely doing something more sophisticated than it seems. They could have a hidden life as computers. For example, as we will discuss, NMDA receptors could be sophisticated coincidence detectors. As another example, the calcium concentration in the presynaptic terminal could integrate the number and frequency of action potentials as a probability. Also, the calcium concentration inside the spine head could be encoding the probability of future change in synaptic strength, as we will see soon. Other ions could also encode mathematical variables that the neuron is computing, using its physical and chemical properties. Second-messenger cascades, as one more example, are likely amplifying inputs and enhancing their signal to noise. Slower biochemical cascades could be measuring a signal and spreading it out in time, essentially computing the integral of an input. And faster molecules, like ion channels, or fast events like the action potential itself, could be computing the derivative of an input.

All these neuronal mechanisms have computational meaning: neurons do math using the physics and chemistry of their components. The actions also tell us that the brain is not just a digital computer but an organic one that is using its own hardware to compute. If that is the case, it may not be useful to try to replicate the brain in digital media, as you would need to use the brain's same hardware, in all its detail, and will need to build another identical brain if you wanted to replicate it effectively. So much for uploading your brain into the cloud!

Spines

Most excitatory terminals end up contacting dendritic spines (figure 3.7). Spines are small protrusions that cover the dendrites of most brain cells in enormous numbers. A typical human pyramidal cell may have a hundred thousand spines, and Purkinje cells even more. Spines have a bulbous head, about one micron in diameter, on which the excitatory presynaptic terminal makes its contact. The spine head is connected to the dendrite through a thin neck, about a hundred nanometers in diameter and less than a micron in length on average, although different spines have varied morphologies.

Spines were first described by Cajal in 1888 (look at his beautiful drawing at the beginning of this lecture), who guessed that they serve to receive synaptic inputs. Electron microscopy in the 1950s confirmed that Cajal was right, adding the twist that practically all excitatory synaptic inputs

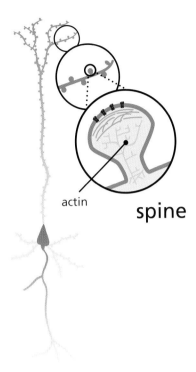

actin

spine

3.7 Dendritic spines are small actin-filled protrusions that cover the dendrites of most neurons in the brain, and receive practically all excitatory synapses.

contact only spines, avoiding the nearby dendritic shaft as if it were cursed. Meanwhile, inhibitory inputs mostly target the shaft, essentially avoiding spines. So nature is parsing the postsynaptic cell into regions specific for glutamatergic or GABAergic synapses.

Why is nature using spines to mediate essentially all excitatory transmission in the brain? Why don't excitatory inputs make synapses on dendritic shafts, which are essentially empty of synapses? Why build spines in the first place, if you have all this available real estate for synapses? The function of spines is still a mystery. They are extremely prevalent, mediating essentially all excitatory transmission, so whatever they do must be crucial in how the brain works. My feeling is that until we understand the spines, we may not understand what the brain does.

There are three main hypotheses about the function of spines. First, by swinging their heads (which they do), they can contact passing axons during the developmental synaptogenesis stage, thus increasing the connectivity of a dendrite. They would be like little arms that shoot out from the dendrite and scan the neighborhood for synaptic terminals and reel them in. This is probably true since spines move a lot during development. Also, spines are filled with actin, a protein whose sole point is to help things move.

The second hypothesis is that spines serve as biochemical compartments. Indeed, when an excitatory input is activated, calcium flows through NMDA receptors into the spine head but remains compartmentalized there. Calcium does not diffuse into the dendrite. This biochemical compartmentalization likely happens with other molecules, so spines could serve as a private molecular reservoir for each excitatory input.

A final hypothesis of the function of spines, not in contradiction with the other two, is that spines are also voltage compartments. That is, they could concentrate higher voltages, generated by the synaptic input, at the head of the spine, but that voltage would be filtered down as it propagates into the dendrite and down to the soma. Voltage compartmentalization could also allow input-specific electrical computations, preventing electrical shunting and interaction between excitatory inputs or the electrical saturation of the dendrite. Voltage compartmentalization could also enable the control of synaptic strength by changing the spine neck's electrical properties.

All these functions could be essential for guaranteeing that neural circuits faithfully work as neural networks, integrating inputs and changing them individually. Indeed, using spines could be the way nature builds neural networks, by endowing its elements with input-specific plasticity, increasing the variety of connectivity and implementing linear integration properties, making spines the anatomical signatures of a neural network. Biochemical and electrical compartmentalization could enable neurons to modify up or down a particular synapse without affecting the rest, providing the mechanisms for input-specific synaptic plasticity. To understand how this could work, let's dig deeper into synapses.

Synaptic Plasticity

Spines have NMDA receptors, which, as you remember from our discussion of neural development, bind glutamate but open only when neurons are active (i.e., fire action potentials) and when, at the same time, they receive synaptic inputs (i.e., released synaptic glutamate binds to them). That means NMDA receptors can serve as molecular coincidence detectors of the input and output of a neuron. When both things happen at the same time—bingo!—sodium and, most important, calcium flow into the spine.

Because of the spine's biochemical compartmentalization, calcium remains in the head of the spine and triggers local biochemistry, which enhances the synapse, making it stronger. That occurs through the insertion of additional glutamate receptors into the spine head as well as shortening and thickening of the spine neck. The bottom line is that the synapse is stronger and better connected to the dendrite (figure 3.8). It has been *potentiated*.

This mechanism, called *long-term potentiation*, or LTP, fits Hebbian plasticity like a glove. So essentially, spines could be the way nature makes

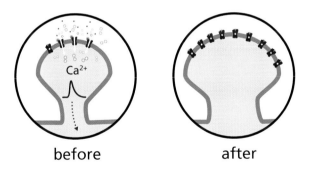

before   after

3.8 Synaptic plasticity is triggered by the influx of calcium into individual spines, if presynaptic and postsynaptic activity spiking occur at the same time. Calcium flows into spines and remains compartmentalized. This can generate increases in glutamate receptors and a stronger electrical connection to the dendrite.

excitatory synapses plastic one at a time to build ensembles of neurons. As Hebb proposed, these ensembles could store memories. But it gets even more interesting. As we discussed, NMDA receptors open only if the postsynaptic neuron is already active when it receives glutamate from the presynaptic cell. This means than some other input has fired that postsynaptic neuron *before*.

How does that happen? It turns out that postsynaptic action potentials propagate into the entire postsynaptic dendritic tree (*backpropagating action potentials*) and invade every spine. But backpropagation takes time: tens of milliseconds. Thus, for bona fide coactivity of the presynaptic and postsynaptic cells to happen at a spine, the postsynaptic cell has to be activated a few milliseconds before the presynaptic cell. When that happens, the NMDA receptors open fully and synaptic plasticity occurs. This requirement whereby the postsynaptic cell fires before the presynaptic cell is called *spike-dependent synaptic plasticity*, or STDP. The temporal properties of STDP, where one neuron (the "post") fires before the other (the "pre") means that the synapses can "associate" the temporal order of events, capturing the causal relation between two stimuli that follow each other in succession, and transform it into stronger synapses and ensembles. So synaptic plasticity can be used to learn the temporal properties of inputs; in this way, the brain can store memories and capture temporal patterns of events in the synaptic connectivity of the circuit. And the temporal patterns of events, as philosophers have been telling us, reflect the causal structure of the world. In Cajal's drawings of neurons with spines, we are looking not only at the hardware of our memories but also at the map of events in our world.

Synapses are changing all the time; they are plastic. The strength of the synapse can change up or down depending on the past history of activity of that neuron and the neurons that are connected to it. But action potentials don't change in amplitude. So why make synapses plastic at all? As we will see, the logic again could be at the circuit level, since plastic connections are exactly what you need to build a neural network that learns. The fact that synaptic plasticity is so prevalent in the nervous system can be interpreted as saying that the nervous system is essentially a learning machine.

Inhibition

If you think about it, this is extraordinary: excitatory circuits capture temporal patterns of inputs and store them in synaptic connectivity with exquisite precision, input by input, spine by spine. Moreover, neurons that release glutamate (i.e., glutamatergic cells) are excitatory, like pyramidal neurons in the cortex for example, and typically have long axons that tend to project very widely, making contact with a neuron here and there but reaching far from where they are located. So these long axons could ensure that there is mixing of information, so these calculations or computations can be done across the brain.

But what do neurons that synthesize GABA, the GABAergic cells, do? And why do their inputs avoid spines? As mentioned, generally, GABA synapses tend to be inhibitory, counteracting the depolarization generated by glutamatergic inputs. But when you look at the axons of GABAergic neurons, you find something interesting: they are mostly local (figure 3.9). In fact, GABAergic cells, which roughly make up about 20 percent of the neurons in the brain, were originally called *short-axon cells* before we knew that they express GABA as their neurotransmitter. That's why they are called "interneurons" as they remain in the middle of the neural circuits. In addition, these axons have dense local connectivity, in some cases contacting practically every single excitatory cell around them. They form a sort of "blanket" of inhibition, as if they were reining in the excitatory cells nearby, preventing them from firing. That could explain why they don't contact spines: inhibition doesn't care about being-input specific and instead affects dendrites with many inputs at once, as if its role was to hit everyone in the neighborhood. In fact, many GABAergic inputs don't even care about dendrites but target the soma directly, or even the axon initial segment, going straight to where the output of the cell gets cooked. And to top it off, some GABAergic neurons just release GABA on the entire tissue, without bothering to

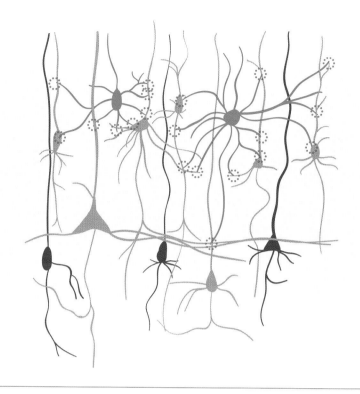

3.9 Inhibition is mediated by GABAergic interneurons that contact most excitatory neurons in their vicinity, extending a blanket of inhibition on the circuit.

make a synapse. So much for specificity! So the strategy of inhibitory inputs is almost the opposite of that of the excitatory inputs. They are not interested in modifying individual inputs or cells; they hit them all, to shut everyone down.

Why do they do that? The function of GABAergic interneurons is still a mystery. One hypothesis is that by counteracting excitatory cells, they could help prevent runaway overexcitation, or epilepsy. Indeed, there is evidence that during normal activation of neural circuits, you open up small holes in this inhibitory blanket. It's as if the brain were built to always be turned off except when needed, which could save energy and prevent unwanted movements or internal states. By the way, this process also echoes what can be seen in other biological networks, like genetic networks, where a large part of the regulation is actually negative. Many genes are repressed, and pathways are activated by removing a repressor. Jacques Monod said that repression is the secret of life, and perhaps his mantra applies to the brain as well.

Another hypothesis is that GABAergic interneurons could help synchronize populations of excitatory cells. That function is a little counterintuitive: imagine that a group of neurons are all firing on their own. Now, if you inhibit everyone at once, when that inhibition ends, as their hyperpolarization wanes, together they could reach the action potential threshold, and all neurons would then fire together. Such synchronization of local excitatory cells is just what you need to get an ensemble started.

A final hypothesis about the function of GABAergic interneurons may have to do with the way activity spreads throughout a circuit. In the same way that action potentials have a refractory period, inhibitory cells, which are often connected to each other through gap junctions, are easily triggered by excitatory neurons, tripping the entire inhibitory blanket and generating a refractory period for the circuit. That ensures that activity gets propagated forward and doesn't get stuck or go back. So inhibition could carve spatiotemporal patterns in the activity of excitatory cells and send the activity of excitatory neurons one way or another, similar to how the switches in a train station can send the train to one destination versus another.

One final thought to explain local and blanket-style inhibition is that it could serve to map inputs onto a circuit far from each other in a computational space, or help ensembles become more distinct from one another, as one ensemble could trip a blanket that would block the other ensembles from firing. This is called orthogonalization and is a property of self-organizing neural networks and could explain why the brain is full of maps, as we mentioned already and will discuss again through the rest of the book.

These hypotheses are not incompatible with one another. They also nicely interlock, just like the hypotheses of spine function. An inhibitory blanket could help prevent epilepsy, trigger ensembles, and carve pathways for circuit activity and separate inputs, and ensembles. If that is true, then the nuts and bolts of computation would be carried out by excitatory neurons, while inhibitory GABAergic interneurons help them get restrained and organized and direct traffic.

Integration

Now let's complete our neuronal trip by looking at what happens to all the synaptic potentials as they are received by dendrites and get integrated at the soma, back where we started. Here, or in the axon initial segment, is where the critical decision is made: to fire or not to fire. That's the entire point of a neuron, to transform its inputs into a digital code of action potential outputs. The basic idea is that EPSPs and IPSPS are summated, and if the intracellular potential happens to reach the threshold to trip enough sodium channels, an action potential is generated and propagates down the axon (figure 3.10). The job is done. If they don't reach the threshold, it's the end of the line. That activity will never propagate beyond that neuron.

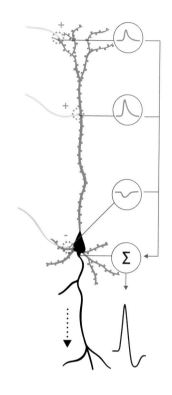

3.10 Excitatory and inhibitory inputs are integrated in dendrites and soma and the final decision, to fire or not to fire, is made at axon initial segment.

The summation of excitatory synaptic potentials by dendritic spines appears to be linear. Inputs are added arithmetically, as when you learn that two plus two equals four. This finding is surprising since essentially all the molecular and cellular machinery in a neuron is built of nonlinear elements. So nature must make a special effort to make things linear. Linear integration makes a lot of sense if you want every input to count at the end, as if you were tallying them in a neuronal "democracy." If spine integration were not linear, larger inputs could dominate, shunting and annihilating smaller ones, and the benefit of having a large connectivity matrix with many inputs would be lost. Thus, the idea of a distributed circuit goes hand in hand with a linear integration regime.

But in cortical neurons smaller pieces of dendrites can also become activated independently, generating dendritic spikes. These local spikes could be based on calcium and sodium channels and NMDA receptors. How would these local spikes square with the linear summation of spines? One can imagine a two-step model, with a linear integration by spines first occurring everywhere in the dendritic tree, and then regions where clusters of spines are particularly active could generate dendritic spikes, which then themselves would interact to fire the soma. Thus, local dendritic spikes could serve as dendritic equivalent to the axonal initial segments and collaborate between them, or perhaps compete among them, to fire the axon. In this view, the dendritic tree would be broken up into dozens of semi-independent segments, and each neuron would be the equivalent of a small circuit.

How about the integration of inhibitory inputs? They could be linearly subtracted from the excitatory inputs, or they could shunt EPSPs, leaking their current—in which case, they would serve as divisors—and even vetoing altogether the response of a neuron. GABAergic inputs, which specifically target dendritic shafts, could also control dendritic spikes and determine which are the lucky segments of the dendrite and the associated clusters of spines that get to dominate the soma. So, just like at the circuit level, inhibition would again serve to regulate excitation at the neuronal level.

Synaptic inputs are small and slow, particularly compared with action potentials, which are at least an order of magnitude larger and faster. Why is that? These characteristics must be important. For example, an EPSP generated by a single spine, when it does not fail, has an amplitude significantly smaller than 1 mV at the soma. That's nothing: you need at least 20 or 30 mV to fire the neuron. That means you need at least dozens (or hundreds, if inhibition is involved) of spines to be active at the same time to reach the action potential threshold. Perhaps local dendritic spikes can help the spines make themselves heard. It also pays for spines to be active in synchrony, so they can magnify their small effects, which implies that neurons could be built to work together as synchronous groups. If they don't, their message will go nowhere; it will die out in the dendrites. Perhaps to help keep the ball rolling, EPSPs are also particularly slow, much slower than the action potentials, which could enhance integration. If you want to integrate—that is, summate—many signals, you have to take your time. In fact, rather than representing the time of the firing of a presynaptic cell, you can think of an EPSP as encoding the probability that the presynaptic cell has fired. In that light, it makes sense that EPSPs are slow.

Neuronal Cell Types

The brain would be simple if all the neurons were identical, but that is not the case. There are different classes of neurons. Glutamate and GABA give us two basic types of neurons: glutamatergic, which are excitatory, and GABAergic, which are inhibitory, as we have discussed. But that doesn't even scratch the surface. It turns out that through a combination of anatomical, physiological, and molecular methods, we are confirming what Cajal already knew: that there are many subtypes of neurons (figure 3.11).

In the primary visual cortex of the mouse, which has been better studied, there are at least one hundred forty classes of neurons! They differ in shape, connectivity, and likely also function. And, as Cajal and Lorente de Nó intuited, most of these neuronal subtypes are GABAergic. So why do we need all these subtypes of neurons, and particularly so many GABAergic ones? A hint may come from the retina, a part of the central nervous system that has been thoroughly mapped anatomically and physiologically.

The retina has dozens of cell types, and each type, with specific morphological, physiological, and molecular characteristics, appears to be exquisitely designed for a particular function. In fact, as is the case in any other part of the body, the retina demonstrates the beautiful marriage of form and function in biology. If such is also the case for the cortex, you

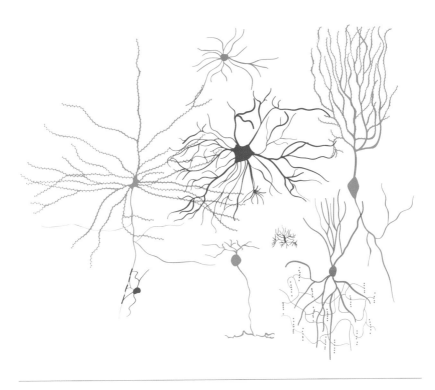

3.11 Excitatory and inhibitory neurons belong to many different **subtypes**, with different dendritic and axonal morphologies, and different roles in the circuit.

could predict that each of these 140 subtypes of neurons may be carrying out a particular kind of computation.

What are those 140 different computations? Each type of glutamatergic neurons could be part of a specific circuit. Meanwhile, each subtype of GABAergic neuron could be tailored to deal with inputs that have particular temporal properties. More generally, neuronal cell types could inject variety into a circuit and enable your network to explore a larger computational space in time and perhaps also in terms of the variety of mathematical operations that the neurons carry out. If all the neurons were the same, your computational options would be limited. From a neural network viewpoint, the presence of cell types would be a great thing, as it would expand and enrich the network's function.

These differences between cell types may be important, and the computations they carry out could be very different from each other, in which case, we need to figure them out one by one. Or perhaps not; perhaps these differences are minor in the larger scheme of things and simple coarse-grained models of circuits with only two types of cells—excitatory and inhibitory—cells will suffice to explain the brain.

Glia

Last but not least, there is a critical type of cell in the brain we haven't yet discussed. Besides neurons, the nervous system is filled with glia, also called neuroglia. These cells are probably as numerous as the neurons themselves, and they come in three main flavors: astrocytes, oligodendrocytes, and microglia. Astrocytes are the most common glia and, like neurons, can also be divided into several subtypes. Astrocytes are small cells with short, exuberant protrusions that wrap around synapses. Their function is still unclear: originally they were thought to provide metabolic or mechanical support for the neurons, but there is increasing evidence that they play an active role in regulating synaptic transmission and, by doing so, controlling the function of ensembles and brain circuits. But astrocytes themselves don't have synapses, and compared with neurons, they act slowly. They interact with neurons indirectly by taking in or secreting ions, metabolites, second messengers, and even neurotransmitters or their precursors. But whatever they do must be essential because there is no brain tissue without astrocytes.

Oligodendrocytes are different from astrocytes, and we have a good idea of their function. These specialized glial cells have membrane protrusions that wrap themselves around axons, insulating them with sheets of membrane that, as we noted, is called *myelin*. As we discussed, they act like the plastic insulators around electrical cables to prevent current from leaking out, except that in the brain, the job is done in piecemeal fashion, and each oligodendrocyte takes care of insulating only a small segment of the axon.

Finally, microglia are thought to serve as macrophages, ambulating between the brain tissue and cleaning up foreign bodies and dying cells and repairing the tissue. Not surprisingly, their role in brain disease and inflammation is important. All glia engulf or wrap neurons, or parts of a neuron, something they have in common. Indeed, the term glia comes from the Greek for "glue."

▼ RECAP

Neurons are a marvel of nature, built with unbelievable precision and great beauty, and, on top of all that, neurons are self-assembled, like all biological systems. They are the ultimate inspiration for nanoscientists. The structural beauty of neurons and the precision in their assembly raises many functional questions: why is the nervous system using action potentials, synapses, and spines, and why does it have all these different classes of neurons? Are there any general principles? The answer is a resounding yes. Not only are there common principles, but they are conserved across neurons, brain regions, and species, confirming the universality of nature. Most neurons, in most brain areas and in most animals, all work essentially the same, as sophisticated nanomachines built for fast transmission of digital signals, which are then transformed by stochastic synapses into synaptic potentials, which are then integrated at the axon initial segment to generate action potentials. Synapses are plastic and constantly changing, depending on the patterns of activity, and so are the circuits they build. Different cell types enrich circuit activity and enable them to build reflexes and carry out computations. All these properties make sense if you consider them from the circuit point of view.

Further Reading

Cornejo, V. H., N. Ofer, and R. Yuste. 2022. "Voltage Compartmentalization in Dendritic Spines in Vivo." *Science* 375: 82–86. Imaging voltage in cortical neurons from a living animal, during spontaneous and evoked sensory activity, shows a rich phenomenology of spine and dendritic activity.

Kandel, E. R., J. D. Koester, S. H. Mack, and S. A. Siegelbaum. 2021. *Principles of Neuroscience*. 6th ed. New York: McGraw-Hill. I recommend this book again. It has an excellent section on cellular neuroscience.

Karnani, M. M., M. Agetsuma, and R. Yuste. 2014. "A Blanket of Inhibition: Functional Inferences from Dense Inhibitory Connectivity." *Current Opinion in Neurobiology* 26C: 96–102. A discussion of the potential logic of inhibitory circuits.

Katz, B. 1966. *Nerve, Muscle and Synapse*. New York: McGraw-Hill. A classic, short monograph on synaptic biophysics describing the exquisite science behind the discovery of stochastic transmission.

Kuffler, S. W., and J. G. Nicholls. 1976. *From Neuron to Brain: A Cellular Approach to the Function of the Nervous System*. Sunderland, MA: Sinauer Associates. An excellent review of neuroscience from a biophysics point of view, looking up at the machine from the bottom.

Markram, H., J. Lubke, M. Frotscher, and B. Sakmann. 1997. "Regulation of Synaptic Efficacy by Coincidence of Postsynaptic Aps and EPSPs." *Science* 275: 213–15. The first description of spike-timing-dependent synaptic plasticity. An exquisite study.

Monod, J. 1972. *Chance and Necessity: An Essay on the Natural Philosophy of Modern Biology*. New York: Vintage. A classic book by a Nobel Prize winner, providing deep insights into biology.

Yuste, R. 2010. *Dendritic Spines*. Cambridge, MA: MIT Press. A monograph in which I try to put together all that we have learned about spines.

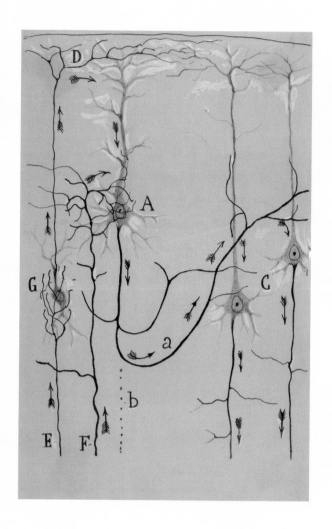

Current pathways through pyramidal cells from cerebral cortex. Armed solely with a microscope and through astute observation, Cajal correctly deduced the flow of electrical activity in many neural circuits. Courtesy of the Cajal Institute, Cajal Legacy, Spanish National Research Council (CSIC), Madrid, Spain.

LECTURE 4: CIRCUITS

But because in my youth, full of pride and somewhat arrogant, I ignored the fear of making mistakes, I ventured ahead into the enterprise . . . in that fearsome jungle, where many investigators have lost themselves.
—Santiago Ramón y Cajal, *Recollections of My Life* (1917)

OVERVIEW

In this lecture, we'll learn about

▶ **The neuron doctrine**
▶ **How central pattern generators work**
▶ **What neuronal ensembles are**

In the last lectures, we learned that the whole point of neurons is to connect to other neurons, forming neural circuits. I also proposed that the reason neurons are built the way they are concerns the role they play in these circuits. The circuits, as we learned, implement the principles of neuroscience that explain how you can use the neural hardware—with hierarchies of ensembles, connectivity, maps, learning, control theory, and optimization—to predict the future and choose an action.

How exactly do neural circuits do all this? As we learned, they are built with hundreds of excitatory and inhibitory cell types using neurotransmitters and neuropeptides, and they extend through the brain—actually, through the entire nervous system. There is likely not a neuron that, through its synaptic connections to other neurons, is not connected in a few steps to all the other neurons in the nervous system. That is a vast network of connections. How vast? Close to one hundred billion neurons with probably around one hundred trillion connections. Just for comparison, the entire internet, when this book was written, was estimated to have thirty billion nodes. So a single human head has three times more neurons than the nodes in the entire internet. No wonder Cajal called neural circuits a "fearsome jungle."

If you are in standing the middle of a jungle, it may be confusing, but from above, things look much more manageable. Things are fearsome when we don't understand them. Let's try to simplify this jungle. At the end of the day, we are dealing with two types of connections: those that can be made from one set of neurons to another in a feedforward, or bucket brigade, fashion, or the connections that are made with your same group of neurons in a recurrent, feedback manner. This simplification may help us understand these jungles, as we will see in this lecture.

We're looking for a framework, a theory that could explain what neural circuits do. This is one of the dreams of neuroscientists: to understand neural circuits. Their structure, algorithms, and functions as they somehow implement representations and predict the future. Neural circuits are the middle ground between neurons on the one end and a behaving animal on the other. Is there such a framework? We touched on this in the first lectures, discussing reflexes and spontaneous activity and ensembles, and now we will unpack that and dig deeper.

Neuroscience, like any science, is inextricably linked to the language and mental models we use to think about it. And scientific progress cannot be understood without important shifts in collective images and language with which we describe the world. Thomas Kuhn, in his book *The Structure of Scientific Revolutions*, describes the history of science in terms of drastic shifts in scientific paradigms which happen in major intellectual battles. So much for science being boring! His proposal was that at any given time in any field of science, there is a widely accepted paradigm, or a theoretical framework, with commonly accepted laws and standards and within which the scientific community operates. Usually paradigms work well, but at some point, researchers begin to notice irregularities or exceptions that cannot explain some of the natural phenomena. Those problems initially tend to get swept under the rug. Yet at some point, people become aware of the growing bumps in the carpet. This is not ignoring momentarily a small obnoxious thing, in a Brennerian fashion, but simply canceling another way of thinking. Once anomalies become too frequent or too hard to ignore, a crisis occurs, and the paradigm is abruptly replaced by a new model that not only explains everything the old model did but also explains the anomalies that the older model couldn't. And the cycle continues. According to Kuhn, that is how science advances, not by a gradual collection of data and knowledge but by crises, or revolutions, in which scientific paradigms are replaced.

Neuron Doctrine

The traditional paradigm of neuroscience is the *neuron doctrine*, which argues that individual neurons are the unit of structure and function in the nervous system. You will find this assertion in the first chapter of every other neuroscience textbook. Such conceptualization of the nervous system, ushered in almost one hundred years ago by Cajal and a contemporary, Charles Sherrington, reflected Rudolf Virchow's cell doctrine, which revolutionized biology in the nineteenth century. They argued that the units of tissues and organisms are the individual cells, and that cells always originate from other cells.

These cell or neuron doctrines, which were the foundation for twentieth-century biology and neuroscience, did not come out of thin air; they directly reflected the methods that these early pioneers used. In particular, microscopes and histological stains like the Golgi method enabled them to study, for the first time, individual cells or individual neurons. At the same time, electrodes made it possible, also for the first time, to record the electrical activity of individual neurons (figure 4.1). It is not surprising that, armed with microscopes, Golgi stains that label individual neurons, and electrodes that record their activity, neuroscientists then proceeded to invest their efforts in describing the structure or functional properties of individual cells in all their glory, taking apart these neural circuits one neuron at a time. It's not a coincidence that Cajal and Sherrington first proposed the neuron doctrine, as they were the pioneers in using these single-neuron methods. Incidentally, its name was a matter of religious belief: it was called not the "neuron hypothesis" but the "neuron doctrine"—a dogma.

Reflexes and Receptive Fields

If single neurons are regarded as structural and functional units of the nervous system, it makes perfect sense to assume that each single unit is special, differing from others by its specific neural connections and functional properties. Cajal spent his life describing these neurons and their connections with the utmost care because he believed that each one mattered, as it had a specific function. In fact, in Cajal's description of his work, there is specificity everywhere: in the dendrites, axons, and connections, an astoundingly complex jungle in which everything is specific and every leaf matters.

This anatomical specificity was music to the ears of the physiologist Sherrington, who formulated the physiological neuron doctrine,

4.1 The **neuron doctrine**, the idea that individual neurons are the structural and functional units of the brain, arose from single neuron methods like the Golgi stain and the electrode.

stating that individual neurons are the unit of function in the nervous system. Sherrington was the first to use electrodes to record the activity of individual neurons. To describe his results, he invented the concept of *receptive fields*—that is, the specific stimulus that a neuron responds to—as a way to measure the function of each neuron. The receptive field is a fundamental concept in neuroscience and has been used effectively ever since. From this perspective, the workings of the brain resemble a gigantic machine in which each neuron responds to the inputs it receives, integrates them, with its receptive field, and then passes the information to the next neuron, which then has a more sophisticated receptive field, because it uses information that has already been "pre-chewed." This process builds an increasingly more selective hierarchy in which every level is more complex than the previous one.

As I alluded to in the first lecture, Sherrington used this hierarchical concept effectively to explain reflexes, such as the knee-jerk reflex (figure 4.2). In his book *The Integrative Action of the Nervous System*, he depicted the brain as an enormous table of reflexes in which neurons are doing nothing until they are stimulated by a specific input, at which point they generate an output. The knee-jerk reflex is simple and easy to understand, but one could extrapolate this idea to all behaviors like running,

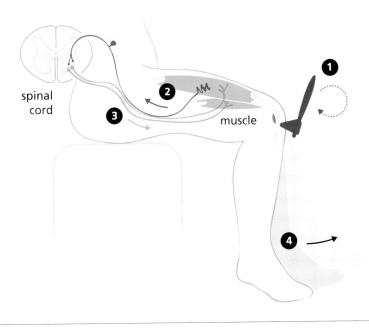

4.2 The **knee-jerk reflex** can be understood as a step-by-step engagement of individual neurons. A hammer hitting the patella (1) stretches the muscle, activating sensory neurons (2) which then activate motor neurons (3), which finally contract the muscle that lifts the leg (4).

brushing your teeth, or even thinking. They could all be nothing more than complicated reflexes, essentially the same process, set in motion by a stimulus and, after much chugging along, generating a motor action. Thus, to understand how the brain works, we just need to describe all its reflexes, collecting them into a long list that translates the transformation of all possible inputs to all possible outputs. Simple: fill in this table of reflexes, and we're done!

Orientation Selectivity

Because of their work, and their neuron doctrine framework, Cajal won the Nobel Prize in 1906 and Sherrington in 1932. Both founded schools of neuroanatomy and neurophysiology that remain influential. In fact, two scientific "great-grandsons" of Sherrington—Torsten Wiesel and David Hubel, whom you know already, and were trained by Kuffler, Eccles, and Sherrington—provided some of the best data to support this view of the brain, in which neuronal receptive fields become increasingly more sophisticated, eventually responding to extremely specific stimuli. To understand their discoveries, we need to explain how the visual system is organized in our brain (figure 4.3).

The visual pathway measures photons in the retina, and retinal neurons project to a nucleus in the thalamus—the lateral geniculate—and its neurons project to the primary visual cortex. If you use electrodes to record the receptive fields of these neurons (i.e., the properties of the visual stimuli that make them fire), as Hubel and Wiesel did, you will find that the retinal neurons have concentric receptive fields (i.e., they fire to stimuli in concentric rings—more on that in lecture 6). Hubel and Wiesel got bored and moved to the lateral geniculate nucleus (LGN), which also has concentric receptive fields. Boring again! But when they recorded from the cortex, they found, serendipitously, that neurons preferred oriented stimuli, like bars of lights. And we are not talking about an odd neuron here or there—practically all neurons in the primary visual cortex responded to oriented bars, with different cells responding to different orientations of the bars.

That was a huge surprise. In centuries of introspection about how we see, no one had ever dreamed that our visual cortex analyzes the world by

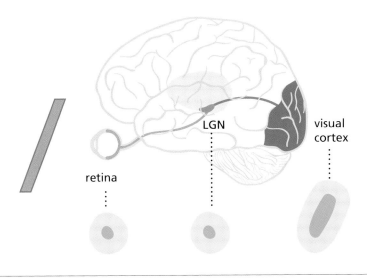

4.3 The **visual system** is organized in a hierarchical fashion, from the retina to the lateral geniculate nucleus (LGN) of the thalamus, to the visual cortex. Receptive fields become increasingly more complex as you move up.

keeping track of the orientation of the visual stimuli, which are broken up into little bars or segments. As we will see, this simple proposal became one of the most influential models in the history of neuroscience. This seminal work opened up the study of the function of the cerebral cortex and earned Hubel and Wiesel the Nobel Prize in 1982.

To explain how cortical neurons responded to oriented bars of light, Hubel and Wiesel followed Sherrington's playbook and proposed a model by which the orientation selectivity of neurons in the primary visual cortex resulted from the integration of previous steps. Specifically, they argued that a group of neurons in the thalamus with aligned concentric receptive fields would project to a neuron in the primary visual cortex (figure 4.4). This cortical neuron would thus gain an oriented receptive field simply because of the collinear receptive field alignment of its inputs. This could explain why cortical neurons only respond if a line with a particular orientation appears in the visual field. Therefore, a new functional property

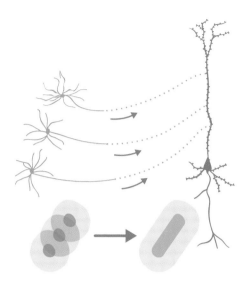

4.4 The **Hubel and Wiesel model** for orientation selectivity explains how oriented receptive fields of visual cortical neurons result from the sum of inputs from neurons with concentric receptive fields, arranged in lines. So the oriented receptive field is shaped like a bar formed by the overlapping sum of the concentric cells.

has been created by the cortex, through the integration of the receptive fields of the neurons below them.

This simple explanation for the generation of oriented receptive fields fits the neuron doctrine perfectly. In fact, the Hubel and Wiesel model was extrapolated to the rest of the visual pathway—and to many other parts of the brain—providing a hierarchical framework to understand how the entire central nervous system works. As Sherrington had imagined, it could be a set of cascades of increasingly more sophisticated receptive fields and functional properties.

The hierarchical processing embedded in the neuron doctrine was extended even to psychology in the 1970s by Horace Barlow. He argued that individual neurons were not only the functional units but also the psychic unit (i.e., the units of perception). He proposed the idea of a grandmother cell; that is, a neuron that has such a specific receptive field that it fires only when you see your grandmother. The grandmother cell is an example of the extreme specificity that neurons could possess, the crown jewel of the neuron doctrine. A decade later, in confirmation of this idea, researchers who recorded higher visual areas in monkeys found neurons that responded to faces. Moreover, these neurons responded to the faces of specific monkeys and not others. Case closed?

Spontaneous Activity

Not so fast. Let's look at what's under the rug. The Sherrington model would predict very specific receptive fields, but the orientation tuning of the neurons that Hubel and Wiesel recorded was actually very broad. These neurons were not very good at measuring orientation. Moreover, to measure these receptive fields, experiments had to be done by averaging responses after repeated presentation of the stimulus and using animals under anesthesia. Otherwise, the variability in neuronal responses made the receptive fields hard to measure. But obviously, the brain does not work well under anesthesia and it also does not need extensive averaging to recognize a stimulus, so there was a worrisome disconnect between the idea of receptive fields and ongoing processing in a normal brain.

But the biggest problem has to do with spontaneous activity. Sherrington viewed the brain as an input–output box in which a specific sensory input triggers a specific output behavior in the form of a reflex. And Hubel and Wiesel, with Barlow, cemented the idea of a hierarchically structured visual pathway to define the way we talk about the brain. But does our brain really operate like a machine that becomes activated only when it receives inputs and produces outputs? Actually, the brain is spontaneously active; that is, continuous ongoing neuronal activity is present in the absence of inputs. That is probably the case for all neural circuits in all animals, from the very simplest cnidarians to the cerebral cortex of humans. And as we will see, this spontaneous activity is not random noise; it has peculiar structures in space and time, with moments when some neurons fire before another set of neurons fire, and so on.

Although the function of spontaneous activity is still under intense investigation, the point is that it is there, and it exists independently of sensory stimulus, independently of the outside world. This is not a small detail: activity that is independent from the sensory world could enable it to build a representation of the world, which can exist and operate on its own. This virtual universe may be what we call the mind.

Let's take a closer look at this spontaneous activity business. According to Kuhn, that is how science should work: we look for exceptions to come up with a better model and switch paradigms if needed.

Central Pattern Generators

The questioning of Sherrington's paradigm started in his own laboratory. Between 1910 and 1915, Sherrington's student Thomas Graham Brown was studying how cats walk and performed an apparently odd experiment. He cut off all inputs to the spinal cord, severing connections from the sensory periphery and the brainstem. He was expecting, like a good Sherringtonian, that stopping all input would stop the machine. But, astonishingly, he found that animals could perform near-normal locomotive movements without any input into the spinal cord or subsequent sensory feedback.

Brown proposed that rather than relying on commands from the senses, the spinal cord has an interconnected network of neurons that operate like a pendulum: they excite each other to produce a spontaneous rhythmic movement, moving the limbs in an alternating pattern. To explain how such a neural pendulum could work in the absence of external inputs, he proposed a *half-center* circuit model, whereby two groups of coactive neurons located on different sides of the spinal cord would inhibit each other. Thus, this cross-inhibition would generate an alternating activity pattern. He coined the concept of *central pattern generators* (or CPGs) as intrinsically active neural circuits with oscillatory properties. Walking was not a reflex, but it was generated internally by the spinal cord. After turning his mentors' world upside down, Brown focused for the rest of his life on climbing mountains, pioneering new routes up the east face of Mont Blanc. What a remarkable person

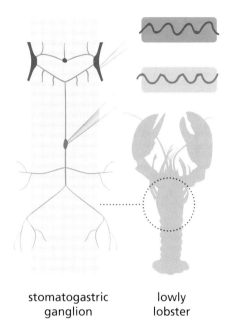

stomatogastric ganglion   **lowly lobster**

4.5 Central pattern generators (CPGs) are neural circuits that generate spontaneous rhythmic activity and exist even in lowly lobsters: their stomatogastric ganglion, through synaptic interactions between its neurons, intrinsically generates a digestive rhythm that chews food.

CPGs are not a curiosity of the human spinal cord but are found in many species. In fact, some of the best-understood CPGs are in the ganglia of crustaceans, in which researchers were able to record from every neuron in a ganglion and explain how the rhythmic CPG operates (figure 4.5). In a set of exquisite experiments, every connection in the lobster stomatogastric ganglion was functionally mapped, and a model of the circuit was built *in silico* that behaves like the real thing. We will pick up the CPG thread in a later lecture, but now let's tackle humans.

Cortical Alpha Waves

Who cares about lobsters or even the spinal cords of cats? Maybe CPGs are used for trivial repetitive motion but have no real place when we try to understand how the "real" brain works and computes. Maybe CPGs are a weird phenomenon we can still sweep under the rug.

Well, the problem didn't go away and the challenges to the neuron doctrine became more serious when electroencephalography (EEG) was invented. During his very first experiment with EEG in 1929, Hans Berger turned to the most readily accessible subject, his fourteen-year-old son. Berger placed electrodes on his son's occipital skull (on top of the visual cortex) and asked the boy to look at an object. That caused strong signals in the EEG, which he happily interpreted as being generated by visual inputs to the brain. However, when Berger asked his son to close his eyes, the EEG, instead of going flat, as would be expected

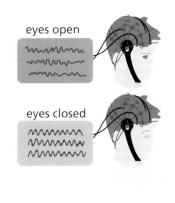

4.6 First EEG recordings. In 1922, Berger measured brain activity from the skull of his son and discovered that the brain is spontaneously active.

from an inactive visual cortex, showed instead robust activity: an oscillatory wave rhythm of about 10 Hz, also known as *alpha rhythm* (figure 4.6). Unexpectedly, the cerebral cortex was active spontaneously, without external stimuli and in a rhythmic fashion, just like the CPGs in lobsters and cats' spinal cords. This was not a fluke: after the discovery of the alpha rhythm, spontaneous rhythms were described in humans and all other animals and in pretty much all parts of the brain.

What is all this spontaneous activity doing? A century later, it is embarrassing to confess that we still don't really understand what the alpha rhythm—or pretty much any oscillatory rhythm—does. But they are always there. When we are awake, most of our brain activity is a cacophony of rhythmic activity of different frequencies, all mixed together. When we are asleep, our brain produces fewer rhythms, but they are stronger. There is only one time when our brain doesn't generate any spontaneous rhythmic activity: when we are in a coma or dead.

Recurrent Chains and Reverberating Activity

The presence of spontaneous rhythmic activity questions our Sherringtonian paradigm at its core. And spontaneous activity is so widespread that you cannot just sweep it under the rug as an oddity. There is a major conceptual problem: What is the point of neurons having receptive fields if they can also be activated without any input? What are neurons really doing? These questions started to poke some serious holes in the neuron doctrine model.

Around the same time that Graham Brown was doing his experiments and demonstrating that his mentor's vision of the brain might be incomplete, another rebel, the Spanish neuroscientist Rafael Lorente de Nó, trained by Cajal, was also contradicting the doctrine stated by his own adviser. Lorente used electric shocks to stimulate neural circuits, and he noticed that neurons often remained activated even after the cessation of input. He called this *reverberating activity*.

Based on those experiments, and careful observations of neuronal connectivity, Lorente began to draw neurons as a part of recurrent circuits, whereby neurons connect back to the same neurons that provided inputs to them. This apparently minute detail marked a defining moment in neuroscience, as it highlighted the importance of the neurons' interconnectivity to do their own thing, as opposed to their slave-style role in a reflex type of action (figure 4.7). Lorente went further to propose that the wiring of most circuits in the brain is recurrent, meaning that neurons are connected to each other through excitatory loops and, as a result, they generated the reverberating activity he had discovered. He called these reverberating circuits of connected neurons *chains* and argued that they were the functional units of the cortex. According to his view, the brain does not simply respond to an input with an output but generates ongoing activity. Lorente's anatomical model could account for the anomalies in the old paradigm, explaining what Brown found in his cat spinal cords or what Berger saw in his son.

A key aspect of Brown's and Lorente's proposal was the shift in focus from the individual neuron to groups of neurons. Neurons were only

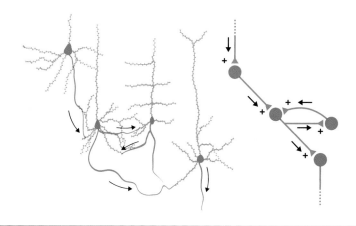

4.7 Lorente proposed that neural circuits are built to generate **reverberating activity** through feedback excitatory connections.

small components of larger functional units. Consider the irony: in the two labs that championed the neuron doctrine, two young disciples provided experimental evidence arguing that both the anatomical and functional units may not be the single neuron but a group of neurons that has its own agenda. Often, the best disciples are those who do the opposite of what they are told. That is one of the wonderful things about science, that it's not hierarchical. To his credit, at the end of his life, Sherrington became more flexible in his thinking and in his Nobel lecture, spoke of the idea of a "neuronal democracy" in which neurons operate as a group. He called these groups of neurons *ensembles*.

Neuronal Assemblies

The next step in the program was to infuse these groups of neurons with some functional meaning. Donald Hebb, whom you may remember from past lectures, took Lorente's reverberating chains a step further. In 1949, he published *The Organization of Behavior*, in which he speculated about how neural circuits would work to generate behavior and memories. Following Lorente, Hebb assumed that the networks are wired in a recurrent

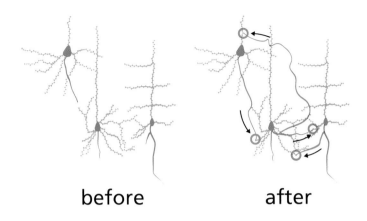

before   after

4.8 Hebb proposed that connected neurons that fire at the same time strengthen their connections, building a **neuronal assembly**, i.e., a small circuit that then becomes activated as a unit and can encode a memory.

fashion so that information can go around in loops. According to Hebb, neurons in such loops would not only fire one after another repeatedly, but, importantly, due to plasticity in their synaptic connections, their activity would reinforce their connections, thus building a group of neurons that work as a unit (figure 4.8). To explain how neuronal activity would change synaptic connections, he proposed the following synaptic plasticity rule, which we call *Hebb's rule*: "When an axon of cell A is near enough to excite a cell B and repeatedly or persistently takes part in firing it, some growth process or metabolic change takes place in one or both cells such that A's efficiency, as one of the cells firing B, is increased" (Hebb, 1949).

Another way to put this is, "neurons that fire together, wire together," as I mentioned in our discussion of learning in the first lecture. As much as he credited Lorente's work, Hebb gave these chains and reverberations a new name: *neuronal assemblies*. His model explained how they could be built naturally in neural circuits, with a simple learning rule. In fact, if you employ Hebb's learning rule, and activity flows through a circuit with some degree of interconnectivity, you cannot avoid building assemblies.

Hebb also gave us an idea of what these groups of neurons could be good for. Given that they form with synchronous activation of presynaptic and postsynaptic cells, assemblies could bind two different inputs, which explains how you can associate two different events into memory. Think of Pavlov's dog. Let's say that input A (bell) drives the first neuron and input B (food), the second neuron. Hebb's rule would predict that after a few repeated presentations of the inputs in the same order, the connections from the first to the second neuron would be strengthened and an assembly would be formed. Afterward, just the first input would trigger the assembly and fire the second neuron, even if there is no food. In other words, the temporal link between both inputs has been stored in the wiring of the circuit. You have built yourself a memory and captured a causal link in the world.

Hebb's proposal meant that neuronal activity could change neural circuits. Thus, the brain is not just an input–output table of reflexes but a living machine that is reacting to the world and changing its physical structure—in other words, a learning machine. This idea, which gets close to the heart of neuroscience, ties together synaptic plasticity with the activity of groups of neurons, which could implement memories, or mental states.

Synfire Chains

Following the work of Lorente and Hebb, many researchers explored the idea that neurons are activated in groups. Perhaps one of the most interesting proposals came from the neurophysiologist Moshe Abeles. Recording the activity of cortical neurons in monkeys, he pondered why cortical circuits are built with such weak and stochastic synapses, and argued that the only activity that could ever propagate through those barren lands is synchronous activity, whereby neurons fire together in a group, because this way they add their effects together (figure 4.9). Moreover, he also pointed out that as the membrane potential of a neuron gets closer to the action potential threshold, the relation between the membrane potential and neuronal spiking becomes steeper, i.e., nonlinear. This means that

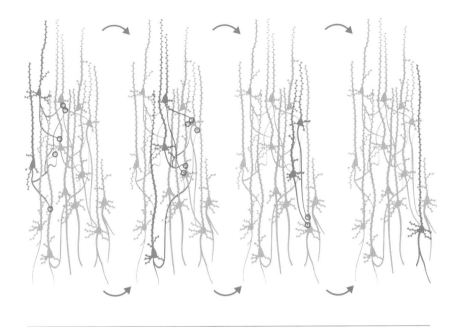

4.9 Abeles proposed that the **synfire chains** are synchronous groups of neurons that activate another synchronous group, propagating neuronal activity through the cortex.

small increments in voltage have a huge influence on the firing. Again, this would select for synchronous activity, as neurons firing together would have a much greater effect on a postsynaptic neuron than when they are firing individually.

From these astute arguments, Abeles proposed that activity in the cerebral cortex propagates via synchronous groups of neurons. Once a group of neurons fires, its activity can propagate only if it manages to activate another group of neurons, and so on. He coined the term *synfire chains* and proposed that these groups of neurons would need to be activated in precise temporal patterns to maintain the tight synchrony of the group. Moreover, if these chains were to become reactivated, they would fire exactly in the same precise temporal order. Abeles and his students found evidence of precise temporal firing in recordings from monkeys engaging in behaviors.

Neuronal Ensembles

Lorente's chains, Hebb's assemblies, and Abeles's synfires are aspects of the same phenomenology: a group of neurons that fire together or in close synchrony. What do we know about these groups? Are they common? Do they have a function? Do they mediate memories? What is the mechanism that builds them and eliminates them? Are they present in different brain circuits? Are they conserved in evolution?

Research depends critically on the methods used. To extrapolate how groups of neurons are firing, researchers originally used electrodes to record the activity of individual neurons. That is not an ideal method if you want to map activity in a neural circuit, since they are built with hundreds or thousands of neurons. This situation changed with the introduction of methods to record from many neurons simultaneously. Using electrical and optical methods, many researchers in the last three decades have described groups of neurons firing in concert, either in synchrony or tight lockstep. Many investigators now use the term "ensemble" to describe a group of neurons firing together, as it is a neutral term that does not adhere to a particular model.

The study of neuronal ensembles is one of the main thrusts of circuit neuroscience and the main focus of my lab, as well as many other groups. Together, we are investigating those and many other questions, and the results are revealing a rich phenomenology suggesting that ensembles are for real: they are common, found in different parts of the brain and species—from cnidarians to mammals—and are involved in functional outcomes.

An example of this work comes from results from our own work using calcium imaging. This method uses optical microscopy to measure the fluorescence from calcium sensors inside neurons. We found (unexpectedly) that there is a strict correspondence between action potential firing and increases in the intracellular free calcium concentration inside the cell. So by monitoring the fluorescence emitted by calcium sensors in a neuronal population, we can detect which neuron is firing and monitor the activity of that population. It is a slow method (you can follow action potentials up to only 40 Hz), but you can see with your own eyes in a video what every neuron is doing. Well, using calcium imaging, when you look at the activity of cortical circuits in an awake-behaving living animal, you see neurons firing together in groups. These synchronous ensembles actually dominate cortical activity. Moreover, synchronous ensembles also occur spontaneously, when the animal is not doing anything. In the case of the visual cortex, you can even measure neuronal ensembles when the animal is sitting in the dark.

In another unexpected discovery, we found that when we use optogenetics to activate individual cells in an ensemble, you can trip the entire thing. This phenomenon, known as *pattern completion*, is similar to what occurs with human memory, in which remembering one part of an experience makes us recall the entire experience. That's what happens at the beginning of Marcel Proust's famous book series, *À la recherche du temps perdu* (*In Search of Lost Time*), in which biting into a pastry unleashes years of memories.

Moreover, we also found that ensembles are not just hanging out there but are involved in behavior. If you activate an ensemble in mouse cortex that corresponds to a particular visual stimulus, the mouse behaves as if he had seen that stimulus (figure 4.10). And if you block the firing of that ensemble, the mouse acts as if he hasn't seen that stimulus even if it's right in front of him. So ensembles appear to be necessary and sufficient for visual perception, perhaps as modular building blocks with which perceptions are built. Experiments from other groups in the hippocampus involving reactivating ensembles (or "engrams," a similar term) suggest that ensembles could also be building blocks of memories. That is all quite in line with Lorente, Hebb, and Abeles.

So what's the big deal about neurons working together as an ensemble? As basic units of brain function, ensembles have many advantages over individual neurons. In other experiments in the mouse visual cortex, we are finding that they can last for weeks and have enhanced precision and robustness when coding for visual stimuli, as opposed to individual neurons, whose responses are more labile, variable, and fleeting. No wonder Hubel and Wiesel needed to average the responses of individual cells.

4.10 When a mouse sees a visual stimulus (top), a **neuronal ensemble** of neurons is active (red). When that ensemble is artificially reactivated (bottom), the mouse behaves as if it saw the stimulus.

But perhaps the key advantage may have to do with pattern completion: the ability to engage all the neurons in an ensemble as a unit, as a block. This property enables ensembles to be deployed as Lego-like modules in the gigantic playground of our brain.

The TV Problem

If the function of the brain is based on ensembles, could we ever understand its function by recording the activity from individual neurons, mapping their receptive fields, as we have been doing since Sherrington? Perhaps that is not such a good idea, as we would miss the ensembles altogether.

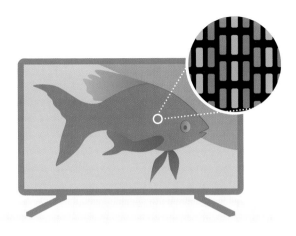

4.11 The TV problem. An image in a screen is an emergent property that arises from the interactions between pixels. Neural circuits likely generate emergent properties through the interaction between neurons.

Consider the innocent task of watching a movie playing on a television screen (figure 4.11). Imagine if you tried to decipher the movie by looking at a single pixel and carefully measuring how it changes under different conditions. That is the exact equivalent of the receptive field experiment. Even if you watched that pixel forever, you would never understand what's going on, because the image on the TV screen is formed by the fleeting interactions of pixels, which fluctuate in space, time, and color.

What if, like the TV, the brain builds "images," or transient states of activities, by activating groups of neurons that would change in space and time? Those images are our ensembles. For example, the primary visual cortex (or V1) of a mouse has about 180,000 neurons. You could try to understand it by recording the activity of one neuron at a time, carefully describing its receptive fields. But you might never be able to figure out its function, since the ensembles become apparent only when you look at the activity of all neurons together as a network. If that is the case, by recording from one individual neuron at a time, we neurobiologists have been essentially trying to watch a movie by measuring one pixel from one TV screen playing one movie and then comparing that with recordings from a different pixel from a different TV screen, perhaps playing a different movie—a hopeless task.

The images on a TV screen are an example of emergent properties of a system: the system has properties that arise from individual elements and thus cannot be described fully through the study of the individual elements. Emergent properties are not magic and do not arise out of thin air; rather, they "emerge" from the interaction of the elements. You cannot capture these interactions by studying the elements in isolation. In fact, that is precisely the definition of an emergent property: whatever cannot be measured by studying the elements of a system in isolation.

Emergent Properties of Neural Circuits

Emergent properties are all around us, not just in every TV screen but also in the physical world and nature. An example of an emergent property in physics is found in a magnet. If you break it into individual atoms, it will lose its magnetism, and you are left wondering where the heck the magnetism went.

Examining how ferromagnetism works is a good example of how emergent properties are not magical but can be understood. In the 1920s, a German student, Ernst Ising, was given as his research topic the "easy" problem of understanding magnetism. Not knowing that this project was supposed to be difficult, he proposed the simple idea that the interactions among atoms in a lattice would generate stable states, known as *spin glasses*, which would be magnetic. This occurred due to atoms interacting with neighboring atoms and behaving synergistically, ending up with the same spins—essentially working together as a unit, just like an ensemble. His model was proven correct: the apparent magic of magnetism is a natural outcome of atomic interactions.

The brain, too, can be viewed as a system of complex networks of particles, the neurons. Do neural circuits have emergent properties? In fact, if you stop and think about it, the brain is ideally designed to generate emergent properties. Any part of the brain is made up of a multitude of neurons, and each neuron is connected to large numbers of other neurons. For example, a typical Purkinje cell in the mammalian cerebellum receives up to 300,000 synapses from individual parallel fibers. A typical neuron in the human cortex may receive about 100,000 inputs from other neurons and connect to 100,000 neurons. Many other parts of the brain are similarly massively interconnected.

If the connectivity matrix of a neural circuit is highly distributed, the functional effect of any given neuron in the system is diluted by the number of connections it makes and receives. In a way, the more connected you are, the less important your voice becomes. In the most extreme case, each neuron would connect to every other neuron in the brain, in which case, taking out one neuron from the pile would not change anything at all. This again sounds like a gigantic democracy, whereby the vote of any particular citizen does not change any election outcome.

Given the immense number of neurons (~100 billion) and connections (~1 trillion), the brain may not just have emergent properties; it may have been designed by evolution precisely to generate them. Our brain could be the mother of all emergent systems! Let's explore emergent properties of neural circuits in the next lecture, which addresses neural networks.

▼ RECAP

You could argue that the entire point of why the brain uses neurons is to build neural circuits. Although we have learned a lot about how neurons work, how neural circuits operate is still very much a mystery. Although we don't yet understand their structure or function, we suspect that, like many things in biology, they are made up of repeated modules, which are conserved in evolution. Neuroscience is switching from the view that each neuron has a specific function to the idea that ensembles of coactive neurons generate functional states. Using ensembles, neural circuits could implement the computational strategies necessary for building powerful models of the world and computationally predicting the future.

Further Reading

Abeles, M. 1991. *Corticonics: Neural Circuits of the Cerebral Cortex*. Cambridge: Cambridge University Press. In this book, Abeles carefully thinks through the biophysical and anatomical contraints of cortical circuits and proposes, in a bottom-up deduction, that activity in the cortex progresses through synfire chains.

Carrillo-Reid, L., S. Han, W. Yang, A. Akrouh, and R. Yuste. 2019. "Controlling Visually Guided Behavior by Holographic Recalling of Cortical Ensembles." *Cell* 178, no. 2: 447–57. https://doi.org/10.1016/j.cell. 2019.05.045. A work from our own lab demonstrating that activating neuronal ensembles can change perceptual states in a mouse.

Carrillo-Reid, L., W. Yang, Y. Bando, D. S. Peterka, and R. Yuste. 2016. "Imprinting and Recalling Cortical Ensembles." *Science* 353: 691–94. A study in which we use optogenetics to build neuronal ensembles with pattern completion properties in the mouse primary visual cortex in vivo.

Hebb, D. O. 1949. *The Organization of Behavior*. New York: Wiley. A classic reference in which Hebb speculates about how neural circuits give rise to behavior, proposing his famous learning rule and the idea of neuronal assemblies.

Kuhn, T. S. 1963. *The Structure of Scientific Revolutions*. Chicago: University of Chicago Press. A rebellious look into the history of science, in which Kuhn argues that it does not advance through staid accumulation of knowledge but by a community of researchers embracing or discarding worldviews.

Marder, E., and R. L. Calabrese. 1996. "Principles of Rhythmic Motor Pattern Generation." *Physiological Reviews* 76: 687–717. A review of a series of systematic experiments in invertebrates that reveal the inner workings of some of the better-understood CPGs.

Shepherd, G. M. 1991. *Foundations of the Neuron Doctrine*. Oxford: Oxford University Press. A scholarly history of the neuron doctrine.

Sherrington, C. S. 1906. *The Integrative Action of the Nervous System*. London: Archibald Constable. The venerable beginning of electrophysiology, ushering in the physiological neuron doctrine.

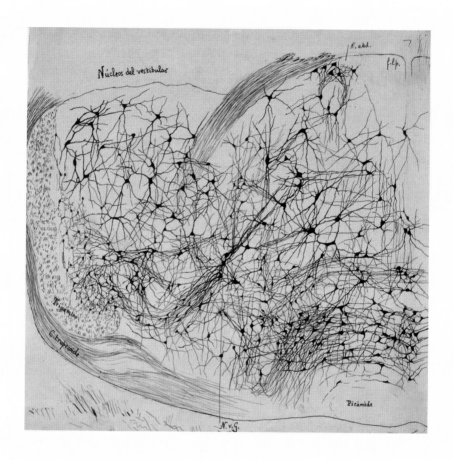

Transversal section through the brainstem of a mouse. This drawing by Cajal's disciple Lorente de Nó demonstrates the recurrent interconnectivity of brain circuits, forming dense neural networks. Courtesy of the Cajal Institute, Cajal Legacy, Spanish National Research Council (CSIC), Madrid, Spain.

LECTURE 5: NETWORKS

And supposing there was a machine, so constructed as to think, feel, and have perception, it might be conceived as increased in size, while keeping the same proportions, so that one might go into it as into a mill. That being so, we should, on examining its interior, find only parts which work one upon another, and never anything by which to explain a perception.

—Gottfried Leibniz, *Monadology* (1714)

OVERVIEW

In this lecture, we'll learn

- ▶ How neural networks work
- ▶ How feedforward networks can abstract
- ▶ How feedback networks generate stable states

Now that we have learned about magnetism, the Ising model, and emergent properties, we are ready to take a fresh look at the impenetrable and fearsome jungles of Cajal. Does the brain generate emergent properties? This question was already posed in the eighteenth century by Leibniz concerning the insides of his imaginary mill: how can the little gears and nuts of the brain generate something as ethereal and comprehensive as the mind (figure 5.1)?

Indeed, the brain generates emergent properties, and they are likely crucial. The study of neural networks can help us understand these emergent properties in a mathematical framework. What are neural networks? They are models and simulations that mimic neural circuits, with interconnected nodes that represent neurons. Since neural networks are theoretical constructs, often detached from real biological data, these models are often called *artificial neural networks*.

Early Neural Network Models

This story begins in 1943, when the neurophysiologist Warren McCulloch and the mathematician Walter Pitts used logical functions and approaches to understand mental processes, effectively creating the first computational models of the brain. They did so by conceiving the most basic neuron, one that can receive any number of inputs

5.1 Leibniz compared the brain to a mill and wondered, if a neuron, like a cog in a wheel, by itself is not intelligent, how could a system built with them generate the mind?

from other neurons, which are denoted mathematically as X_1, X_2, and so on (figure 5.2). Neurons can either be on (and have a numerical value of one) or off (with a value of zero). Each input has a synapse with a particular strength, or a weight (w), which is indexed by the neuron it comes from (w_1 from neuron X_1, for example) and multiplied by the value of the input neuron (i.e., a zero or a one) to determine its overall effect on the receiving neuron.

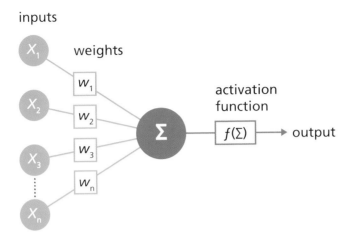

inputs

weights

activation
function

$f(\Sigma)$ → output

5.2 McCulloch and Pitts modeled neurons as units that add up their inputs and, depending on whether the sum reaches a threshold, fire or not.

So, for example, if an input neuron is off, its value is zero, and the effect of all its synapses is also zero. But if a neuron is on, its value is one, and the effect of each of its synapses is a positive number equal to its weight (times one). With this simple scheme, all the inputs received by a neuron can then be summed at the receiving neuron's soma, represented by a circle, and we can easily calculate the total integrated input that a neuron receives. So far, so good. But now comes the rub, the "activation function." If the sum is smaller than a given threshold, the activity of the receiving neuron is set to zero. But if it is above threshold, it is set to one, and that value becomes the new state of the neuron, which is itself also an input to the next neuron down the line and adds its weights to it.

Sounds like a trivial toy, no? Well, as we will explore soon, this toy model makes a neuron into a mathematical function: an integral sum followed by a threshold. That turns neurons into Boolean logical devices, where a one signifies "true" and a zero, "false." So what is the big deal? Well, according to Alan Turing, if you can do Boolean logic, you can compute anything that has a solution if given enough time. These circuits with simple neurons can perform logical computations, like digital computers. These innocent toy circuits are actually universal computers.

Connectivity Matrixes and Learning Rules

The McCulloch and Pitts neuron marked the birth of the field of neural networks. Theoreticians began building more sophisticated neural circuits with many presynaptic neurons connected to many postsynaptic neurons. The connectivity of a neural network can be naturally captured in a *connectivity matrix*, with a series of rows and columns where each row and column represents all the weights of one particular neuron (figure 5.3). Thus, each element in a matrix is the synaptic weight of a connection between two neurons, whereby the element w_{ij} represents the connection from neuron i to neuron j. If the connection is excitatory, the value of w_{ij} is positive, but if it is inhibitory, the value is negative. If two neurons are not connected, the value is zero. This simple matrix elegantly captures, in a standard and powerful mathematical form, all the connections in a

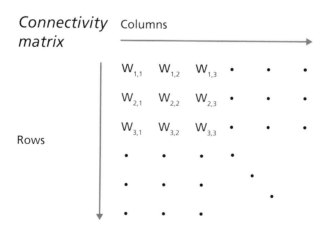

5.3 Connections in a neural network can be represented as a **connectivity matrix**, where each row and column correspond to each of the neurons and each matrix element is the strength of the connection between a particular pair of pre- and postsynaptic neurons.

neural circuit or in an entire brain. You can turn all the jungles of Cajal into a simple matrix.

But having a connectivity matrix is not enough: for a neural network to be able to do anything interesting, the connections have to be able to change. In fact, the connections within a neural network are not fixed—that's the critical point. You need learning. The synaptic weights change according to an algorithm called a *learning rule*. That function captures the activity of a presynaptic or postsynaptic neuron and changes the synaptic strength (or weight) up or down. This modification in weight controls the influence of a particular synapse on the operation of the network. Thus, these toy models are changing one little synapse at a time, and by doing so, they are learning.

Learning rules are normally applied to update all the synapses of the network at every time step. They come in many flavors, and you already know well the most famous one: Hebb's learning rule. We can now redescribe it mathematically with our new nomenclature: when neurons i and j fire together, w_{ij} increases its value.

Armed with a connectivity matrix and a learning rule, and following the McCulloch and Pitts model, we can then apply the activation function to calculate the state of each neuron at any given time (figure 5.4).

The state of the neuron, i.e., how much it fires at time "$t+1$"

Firing of presynaptic neuron at time "t"

Threshold

$$X_i(t+1) = \left(\sum_{j=1}^{n} w_{ij} x_j(t) - \Theta_i \right)$$

The weight of "j to i" connection

5.4 The **activation function** of a neuron in a neural network is an equation that determines whether a neuron fires or not.

As long as you know the state of a neuron at time t, the values of the weight of its inputs, and the states of the neurons that generate those inputs, you can run the equation and predict the state of that neuron at the next time, $t + 1$.

You continue to apply this formula on all the neurons, updating the synaptic weight matrix according to the learning rule at every time step. That way you can calculate numerically how a neural network evolves no matter how complex and large. You just built yourself a learning machine.

Feedforward Neural Networks: Perceptrons

So what can you do with these neural networks? Let's connect a bunch of neurons together and see what happens. Echoing what happens in neural circuits, there are two ways to connect neurons, making two types of neural networks. If the wiring looks like a branching tree—that is, with each neuron linking with the next one in steps—and this stepwise progression is repeated down the line, you have what is called a *feedforward network* (figure 5.5). Feedforward networks have layers with

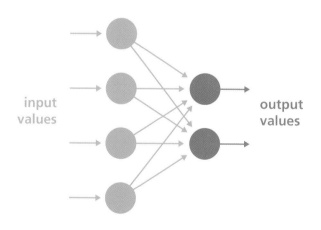

5.5 In **feedforward networks**, inputs (left arrows) activate an input layer of neurons (green) that then connect and activate an output layer (blue).

inputs and outputs, but the outputs never go back to the input layer. It's a pass-through.

The other way to wire up a circuit is by forming loops; that is, when connections from one layer circle back to it either directly or indirectly via another layer. This is referred to as a *feedback* or *recurrent network* (we'll come back to this later in the lecture). These apparently small details in wiring generate major differences in the way networks operate. In the feedforward case, the system is like a bucket brigade and can be decomposed to individual steps, which are easy to analyze. In the feedback network, however, the network can become active on its own, and generate its own states. The way it operates is much more complex but also perhaps more interesting.

Now let's apply our newly learned lingo to the McCulloch and Pitts model. It's a feedforward network with only two layers: an input and an output layer. In further models, additional layers (called *hidden layers*, since they are between input and output) were added to these feedforward networks, and multilayered networks were created (figure 5.6).

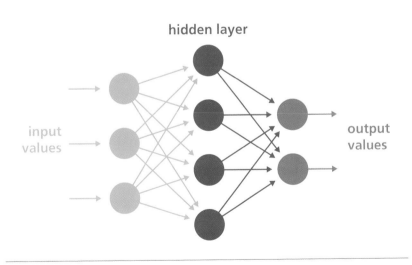

hidden layer

input values

output values

5.6 A multilayer **perceptron** has, in addition to an input and output layer, also a "hidden" layer in the middle.

These models, which could also have positive or negative weights, came to be more sophisticated than the McCulloch and Pitts model and became known as *perceptrons*.

Building a Zip Code Reader

What are perceptrons good for? Why should we care about them? Using the simple learning rules I discussed earlier, you can train perceptrons to perform classification tasks. For example, you can make a multilayer perceptron for digit recognition. When you write a letter by hand and it goes to the post office, it needs to be sorted according to zip code. Back in the day, post offices had designated staff who would read every zip code on all letters to determine where to send them. That can be pretty challenging. (Take it from me, as someone who has to decipher handwriting on written exams.)

You can build an automated zip code reader with a camera to take a picture of the handwritten zip code and then feed every pixel into a multilayer perceptron. But to do that you first have to play a little trick with the thresholds of a typical neuron in the perceptron. As we learned, neurons in a neural network sum the inputs they receive and then become activated and fire, but only if the sum of the inputs is larger than a particular threshold. What happens if we set this threshold high? In that case, the neuron will become active only if all or most of its inputs are on (figure 5.7). Now let's set the threshold low. The neuron now fires every time one of its inputs is on (figure 5.8). These simple cases are more important than they seem: they implement two Boolean logical operations, or "gates," which are of two types:

- ▶ conjunction = AND gate (when two or more input neurons need to be active for a receiving neuron to turn on), or
- ▶ disjunction = OR gate (when any input neuron can turn on the receiving neuron).

So why does it matter that setting the threshold high or low can turn neurons into AND or OR logical gates? Let's say we are trying to recognize

high threshold / **"AND"**

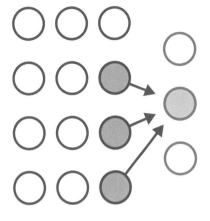

5.7 A network can compute the **conjunction** of inputs (a logical AND gate), if to activate the output neuron (right) one needs the simultaneous activation of several input neurons (left).

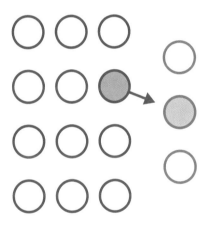

low threshold / **"OR"**

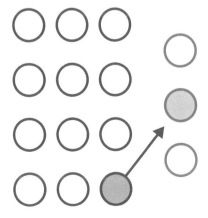

5.8 A **disjunction** (logical OR gate) can be computed if any one of the input neuron suffices to activate the output neuron.

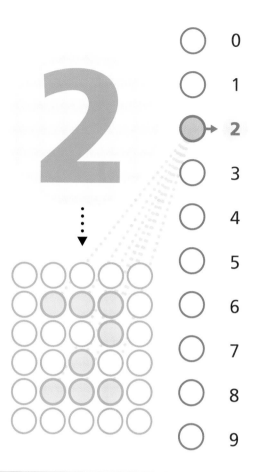

5.9 A perceptron can **recognize digits** when the input neurons that "see" a part of the character are activated simultaneously and they turn on an output neuron, which detects the character in the image.

the handwritten number 2 (figure 5.9). The camera takes a picture of it and sends it to the first layer, the input layer of the perceptron, which measures every pixel of the image, turning into a value of a particular input neuron. Thus, the input neurons that "see" the different parts of the character 2 are activated. This input layer then projects to a second layer, which has high-threshold neurons. Some of the neurons in this second layer will turn on only if they are activated simultaneously by the neurons from the first layer that are seeing the different parts of the number. If you wire the system so that every neuron from the first layer that sees a pixel connected to a neuron in the second layer that fires only if all of its inputs are active, you've just built yourself a number 2 detector!

Now imagine doing this for all the digits one at a time. If we show the perceptron a different number—say, the number 3—we will not activate this 2 neuron in the second layer; a different combination of input neurons will activate a number 3 detector.

Let's get mischievous. What if a person with different handwriting writes the number 2? Our number 2 detector neuron will not work, because there will be small differences in the exact shape of the digit, which will activate a different set of pixels. That is where disjunction (OR gate) comes to the rescue. We can solve the problem by adding another layer to the perceptron, a third layer that becomes the new output layer and receives inputs from the previous output layer, which now becomes the hidden layer. But in our new output layer, the thresholds for the neurons are set low, so they become OR units.

Imagine now that in the second layer, in addition to our original number 2 detector neuron, we also have detectors for many types of number 2s, which get activated when they see many handwritten versions of the number. Now let's wire all the number 2 units from the second layer to one mega-unit "2" in the third layer of the perceptron, which, as an OR gate, will become activated by any possible version of number 2 regardless of the way it was written. Now we have a universal number 2 detector, which will work for many different styles of handwriting. How about the other nine digits in a zip code? We simply repeat the process nine times, building detectors for all the different versions of all the digits in the second layer with our AND gates and connect them to mega-detectors of the digits with our OR gates. We have a zip code reader!

If we think more carefully about how our little zip code reader works, we realize that it's all in the wiring. The reason neurons in the second and third layers respond to digits in different ways is because they are specifically wired to do that specific job. Who wired them? That is where the magic of neural networks is found: nobody did. There was no human designer soldering wires behind the scenes. The neural network was originally randomly wired and trained by applying a learning rule to every connection and making the learning rule enforce the correct response of the mega-detector units. That can be achieved by changing the weights one by one and rewarding them, or strengthening those changes that make the network perform better. But that happens automatically; the circuit is self-assembling via learning.

So by measuring the output of the network and updating the synaptic weights with a learning rule, we can craft a connectivity matrix that will generate a desired output. This process happens little by little. Like a potter molding a mass of clay, the learning rule shapes an initially multipurpose network into a zip code reader. The plasticity of the synaptic weights makes this possible, and the function of the network—zip code reading—becomes an emergent property of the activity and connections of all its neurons, not just a few. For any digit to be properly recognized, we need to know not just which neurons are on but also which are off. By the way, doesn't this remind you of the idea we discussed about what GABAergic inhibitory neurons do? It could be critical for the brain to shut things off. I am also reminded of the Argentinian writer Jorge Luis Borges, who argued that true intelligence comes from observing not only what happens but also what doesn't.

Deep Neural Networks

Multilayer perceptrons are essentially combinations of layers of conjunctions and disjunctions. Conjunctions give neurons *selectivity*: they respond only to a specific stimulus. Meanwhile, disjunctions give neurons *invariance*: they respond to many versions of a stimulus. Multiple layers allow neurons to first respond to a part of a digit, then the entire digit, and eventually different versions of the digit regardless of how it is written. Every neuron acts as a Boolean unit, and these neurons are simple in their function. But imagine extending this process with several additional layers. By adding more and more layers, combinations of AND and OR neurons, we build an increasingly sophisticated and powerful machine, one that can selectively detect and classify different kinds of inputs regardless of the conditions. In doing so, perceptrons can abstract concepts, recognizing and extracting their essence from a cacophony of different versions on an input. This is really powerful!

Since their simple beginnings, perceptrons have become much more sophisticated. In fact, they are currently the basis of the deep neural networks that are becoming the workhorse of the tech industry, in companies like Google, Apple, Microsoft, and Meta, among others. The adjective "deep" refers to their large number of layers. They are also known as *convolutional networks* (figure 5.10). Deep neural networks are operating inside our mobile phones and are used in many applications. For example, Facebook uses them for face recognition, which is more complicated than digit recognition but is essentially a similar problem: the same conjunction and disjunction functions over and over again (Facebook's facial recognition has seventeen layers). Every time you take a picture and upload it to Facebook, the app sends it to a deep learning network, supposedly hosted on a powerful server in Sweden above the Arctic Circle so it gets cooled by the cold air. There the picture runs through a neural network, which detects the face of every person in the database.

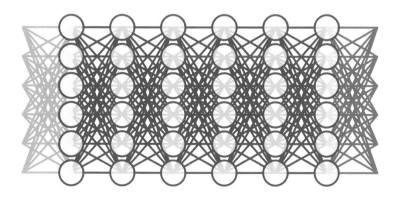

5.10 Deep neural networks are perceptrons with many hidden layers and are becoming the workhouse of the tech industry.

The algorithm then sends the information back and labels the people in the photo you just took. Today, facial recognition networks have better accuracy than the average human. And that is just a taste of what neural networks will be able to do!

The Visual System as a Perceptron

Now let's go back to neuroscience and think about the Hubel and Wiesel model, which you learned about in last lecture 4 (figure 4.4). Guess what? It's a multilayer perceptron! If a single thalamic input cell fires, the cortical cell doesn't produce an action potential, but if the three inputs that respond to an aligned stimulus in a row all fire at the same time, then the simple cells fire. That is a typical high-threshold AND gate.

It gets better. It turns out that when Hubel and Wiesel recorded neurons in the visual cortex, in addition to these types of neurons, which responded to oriented bars in a particular position (the authors called them *simple cells*), they discovered a second class of neurons, which they called *complex neurons*. These complex neurons also responded to visual stimuli that had a particular orientation, but the oriented bar could be anywhere in the visual field (figure 5.11). You could move the bar, and the complex cell would still respond to it. How interesting! How can we explain this? Hubel and Wiesel argued that complex cells receive inputs from simple cells, but if any of the simple cells fired, the complex cell would also fire—a typical OR gate! That explains why the complex cell fires whether or not the input is positioned in the left, middle, or right visual field. A low-threshold third layer in a perceptron can represent this model mathematically.

There is a perfect match between the Hubel and Wiesel model and perceptrons. Mathematically, the famous model is simply a multilayer perceptron with conjunction and disjunction layers, a special case of a feedforward neural network (figure 5.12). In fact, just like in our zip code reader, the conjunction and disjunction layers sequentially implement two fundamental properties for vision: *selectivity* and *invariance*. Perceptual selectivity—that is, the response specificity to a single stimulus—is

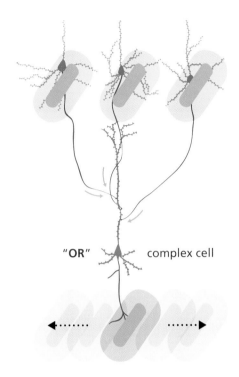

"OR"   complex cell

5.11 Hubel and Wiesel **complex** cells detect oriented bars in different positions of the visual field, because they have low threshold and can be activated by any one of the simple cells (top) that connect to them.

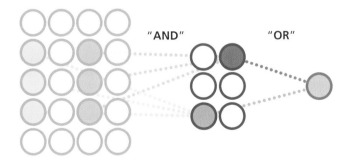

"AND"   "OR"

5.12 The **Hubel and Wiesel model** of simple and complex cells is a three-layer perceptron with conjunction (AND) and disjunction (OR) layers.

critical for detecting objects. And invariance, the ability to detect the same object regardless of the viewpoint, is also crucial for visual perception, otherwise, every time we see an object from a different viewing angle, we would think it's a different object. For example, once you've recognized an object, like the digit 2 (Figure 5.13), you will continue perceiving it as such no matter which way you turn it or whether or not it's moving. That is evolutionarily important because animals move around all the time (and on top of that, we also move our eyes), so it is critical to have a consistent picture of the world to make their behavior coherent and meaningful.

The natural match between feed-forward neural network models like perceptrons and the Hubel and Wiesel model allows us to look at our brains in a different light compared with that of the neuron doctrine. Our visual system could be a gigantic perceptron in which the different steps—from the retina, to the thalamus, to the primary cortex, and so on—are merely perceptron layers. So the visual system, the paradigm of the neuron doctrine, could be a picture-perfect neural network. A critical difference between the neuron doctrine and the neural network model is that the network as a whole is doing the computation, not the single neuron. The entire network is needed. In a perceptron, the neurons that don't fire are as important as the ones that do, as they determine the selectivity. If you

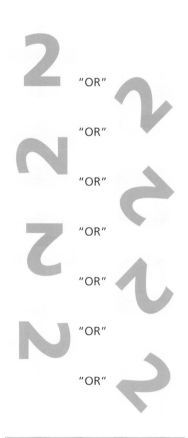

5.13 Perceptual **invariance** can be computed by using disjunction functions for each of the possible views of an object, so the network can detect it regardless of its orientation in space.

consider that, you can reinterpret the receptive fields from the visual cortex as the activation function of hidden layer units within a perceptron.

Using a neural network framework, researchers have built artificial perceptrons to perform all kinds of visual detection tasks. You can extrapolate the model and keep going up the ladder of increasingly more sophisticated layers to imagine how the famous grandmother cells (i.e., neurons that are selective for your grandmother) can be built. So those neurons, the paradigmatic example of the neuron doctrine, end up being simply units from a particular layer in a large neural network.

Feedback Neural Networks

According to Kuhn, if neural networks are the new paradigm, they should be able to explain everything the neuron doctrine model did but also things it didn't. Well, how about that spontaneous activity? We've discussed how the Hubel and Wiesel model is represented through conjunctions and disjunctions within the network, which creates low-threshold and high-threshold neurons. However, the Facebook facial recognition algorithms remain silent if they do not receive any inputs. But we know from our last lecture that spontaneous activity occurs all over the brain. In fact, there is probably no single brain area in any animal that doesn't show a significant amount of spontaneous activity. How can neural network models explain spontaneous activity? One solution to this problem is to consider a different architecture: feedback wiring. By injecting activity back into the system, you can keep it active. To explore that in more detail, we need to bring in two concepts I discussed earlier in the course: ensembles and magnetism.

The Hopfield Model

Perhaps one of the most important steps in the development of the field of neural networks occurred in a 1982 paper by John Hopfield, in which he provided a mathematical model for feedback neural networks and explored the types of activity these networks could generate. Hopfield predicted that feedback neural networks would have ongoing spontaneous activity and also that this activity would settle on stable states, called *attractors*. An example of an attractor is a group of neurons firing together. Haven't we seen this before? Yes, that is exactly a neuronal ensemble! But now we are thinking of them as states in a recurrent neural network.

Hopfield's model was a *completely recurrent network*, which means that all the neurons were connected to all the other neurons (contrary to the feedforward network, in which the neurons in a layer are connected only to the neurons in the next layer). Moreover, the connectivity matrix was *symmetrical*; that is, the connection weight from neuron A to B was the same as from neuron B to A. Under those special circumstances, which are not biologically plausible but can be thought of as an extreme case, the math checked out, and the model was formally identical to the Ising model of magnetism. Just as an atom has a positive or negative spin, neurons can be ON or OFF (they either fire or don't fire), and the synaptic connections between neurons in a network are formally like the coupling coefficients of atoms in a magnet.

Following Ising, Hopfield introduced the idea of the *energy* of a network (figure 5.14). So why do I torment you with yet another equation

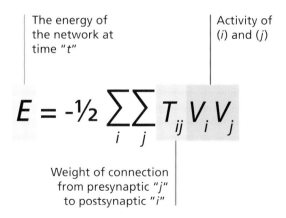

The energy of the network at time "t"

Activity of (i) and (j)

$$E = -\tfrac{1}{2} \sum_i \sum_j T_{ij} V_i V_j$$

Weight of connection from presynaptic "j" to postsynaptic "i"

5.14 Hopfield's **energy** equation calculates the propensity of a network to settle on an attractor, a stable state of low energy, where E = energy, T = synaptic weights and V = neuronal activity. Indexes i and j denote post- and presynaptic neuron, respectively.

to memorize? You don't have to memorize it, you already know it. Do you remember kinetic energy? The formula for kinetic energy is mathematically the same: one half the mass times velocity squared. Is that a coincidence? No. Hopfield used it to conceptually apply kinematics to the analysis of network activity. In other words, the brain according to Newton. That is powerful, because it lets us use our intuition about mechanics to understand computations.

Mass is now the synaptic weights and velocity is the activity of the presynaptic and postsynaptic neuron. Integrating across all neurons, at any given time, you can precisely calculate the overall energy state of the network—that is, the network's propensity for change, which is what the term "energy" means in physics. If the energy is high, the system will change a lot. If the energy is low, the system will not change.

Using this simple equation, Hopfield proved that networks built in this manner will follow an energy function and tend to settle at the lowest energy potential positions in their energy landscape (as when a ball that's rolling through a landscape settles on the lowest point). But notice that we are not talking about physical movement here; we are talking about the propensity of a network activity to change depending on the amount of energy it has. This landscape is the activity landscape of a network; in other words, the landscape of all the possible states of the network.

Looking at the image of network activity as a physical map is very helpful (figure 5.15). In this map, we see in one single view all the potential activity states of a network. Each point in the landscape represents one particular state: for example, one corner could be when all neurons are ON and another corner could represent when all neurons are OFF. Let's use Hopfield's formula and calculate its energy for each of the points (i.e., each state of the network). Now we can add our energy term as a third dimension to our map, or a depth, so the map becomes a landscape where all possible network activity is represented as an uneven landscape with multiple hills and valleys. The valleys represent the lowest energy states. (You'll recall from physics that any system will always move to attain the lowest possible energy state.) Hopfield called these valleys attractors because they attract the activity of the network. These attractors are

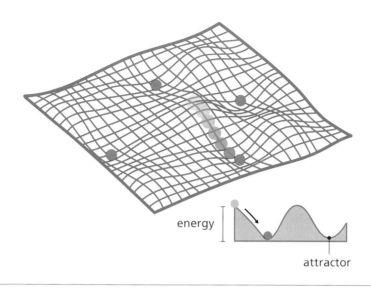

5.15 In Hopfield's **attractor** landscape, each point represents a possible state of activity of the network and the height is its energy. A ball rolling represents how the activity of the network evolves in time (its "dynamics"), and ends up in the valleys (the "attractors"), where the activity settles. Attractors represent a memory, a thought, or a solution to a computation.

stable, or semistable, and Hopfield suggested that they could represent memories or solutions to a particular computation.

For example, imagine that a particular attractor exists in which some of the neurons are on and some are off. The network will stay in that particular attractor until an outside input kicks it to a new point in the landscape. Then the network activity evolves in time, represented in our model by the ball moving around, and the trajectory created by the movement through the landscape would represent the computation itself. Once the network settles, it would end in an attractor, which could be the "solution," or end result of the computation. So the landscape of the activity of the network becomes a map of all of its computations. Isn't that elegant!

As in Newtonian kinematics, the movement of the network through this landscape of activity is called *dynamics*. Not only can it explain the spontaneous activity, which can occur without anyone stimulating the network, but the dynamics become the computation itself. The spontaneous activity is not the noise in the machine but its computational thread! This way of thinking can help explain how some features of human memory work.

Let's go back to pattern completion, which I introduced in the last lecture. Remember Proust? I think you've probably experienced the feeling of struggling to remember some piece of information. But sometimes a simple idea or detail can trigger an entire cascade of memories to flood your mind, bringing back the events or ideas you thought were lost. That feeling of memory unraveling is easy to understand and visualize with attractors. Imagine your memory or brain activity as a ball rolling around the Hopfield network while you are struggling to recollect a memory. Finally, the ball reaches a tipping point at the lip of an attractor and starts falling inside it, toward a more stable state. Every step down the hill triggers a more complete pattern. At that moment, your memory has been triggered, and you've recovered a memory, thus reaching a stable, low-energy point in the system.

This process of full reactivation based on partial input (i.e., pattern completion) can be easily wired in a neural circuit with a small group of interconnected neurons following Hebbian learning (figure 5.16). Pattern

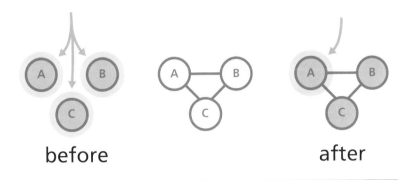

before   **after**

5.16 Pattern completion can be imprinted in a network via synaptic plasticity. If three neurons are simultaneously activated several times (left), this strengthens their connections (middle). After this, the activation of only one of the neurons can trip the entire group (right), completing the "pattern."

completion is a property not just of memory but of many motor behaviors and mental states and can be easily explained by recurrent networks like Hopfield's.

People have criticized the Hopfield model as contrived, because it assumes that the connectivity is symmetric and complete. As far as we know, these two conditions do not hold true for any part of the brain or any nervous system. Still, under those special conditions, you can explore the networks mathematically and prove that they have intrinsic dynamics, attractors, and pattern completion. But just because biological networks are not exactly complete or symmetric doesn't mean that attractors don't exist; it's just that you can't easily explore these attractor landscapes mathematically. The landscapes of incomplete or nonsymmetric networks may not be as clean, but recurrent feedback circuits should still have attractors and intrinsic dynamics.

Here's a final note on Hopfield. His 1982 paper had the word "emergent" in its title. That was not a coincidence. In this model, the computation is not done by any particular neuron, because each attractor is the combination of activity of all neurons—it's a multidimensional structure that exists in time. We have moved a long way from the neuron doctrine. The attractors are formally equivalent to the ensembles, and the Hopfield model explains ensembles and spontaneous activity. Now it all makes sense. The brain is a gigantic neural network.

Beyond Hopfield

Since Hopfield's influential 1982 paper, many researchers have built artificial neural networks with increasingly more complex structures and properties, making neural networks—and computational neuroscience in general—a vibrant area of research today. Because of their impact on artificial intelligence and the tech industry, neural networks are also becoming central to not just neuroscience but also computer science. The relatively stable point attractors of Hopfield have given rise to a new generation of attractor models that can be topologically visualized as lines or rings as opposed to points. Moreover, there are also recurrent neural networks that don't have stable attractors at all; those are more fluid and may be better understood as a continuous dynamics system without stable states. These networks are used as models of many different computations, from general roles as universal "function approximators," which can capture and reproduce any mathematical function, to computational models of memory states, navigational maps, perceptual experiences, probability calculations, prior assumptions, decision making, motor patterns, and control systems for engineering and generators of temporal patterns, among others. The common theme of all these models is their dynamic structure, in which the time evolution of the activity is key.

Although it is fair to say that these models are often way ahead of the experimental data, this exciting field of research has infused modern neuroscience with a strong physics-based theoretical bent. A new field, computational neuroscience, has been born. Neuroscience is starting to resemble many fields of physics, whereby theorists first scout the territory ahead using math and computer simulations and make predictions, which experimentalists then do or don't confirm.

Testing Neural Network Models

So at the end of the day, is the brain just a big neural network? Are neural networks really the new paradigm for neuroscience? Are they for real? Neural networks can explain receptive fields and the Hubel and Wiesel model and, at the same time, account for the existence of ensembles and spontaneous activity, solving the Berger "problem," and giving it a computational interpretation. So they are definitely useful conceptually in our research. The key point is that, in neural networks, the function is distributed: a unit is not a neuron; instead, the group of neurons is the unit. Accordingly, as a classic case of emergent properties, the brain may be better understood in terms of the interaction of neurons in a network rather than by taking into account the individual activity of individual neurons. A new unit of measurement is thus the neuronal ensemble, or a group of neurons, and their distributed interconnectivity diminishes the importance of a single neuron.

In addition to being emergent, if neural circuits work as neural networks, their function could be general: they could be universal computers. You could build a network to do all kinds of things. Also, the same circuit could have many attractors and be involved in completely different computations. Using learning rules, these neural networks would have an ingrained plasticity, which not only forms and shapes them during a training phase but also allows the computations to change. In fact, it would make sense for neural circuits to try to maximize their plasticity, as that is computationally advantageous. Indeed, if you have hardware that is very specific, you will not be able to make connections between neurons if these connections are not already programmed into the hardware. Such specificity by definition limits your possibilities.

With neural networks, in which neurons are widely connected, you maximize plasticity and can associate two things that evolution has perhaps never seen before. Neural networks could explain why mammals have had an upper hand in evolution—the cortex is arguably the largest of the neural networks in the brain; we are essentially cortical animals, and the cortex could learn anything. Moreover, a neural network model can help us understand brain evolution, as a human cortical area that was originally used for actions such as speech may have evolved to also enable us to read and write. They can also help us relate to the ideas presented in the first lecture: maybe the cortex is a machine that generates spontaneous activity to predict the future, and our thoughts are the product of such computations

In this lecture, I have provided a brief summary of the history of artificial neural network models, but these are still early days. A critic would argue that there are no clear cases in which a biologically inspired neural network model has been completely validated. Many exciting areas of research are using methods to simultaneously measure the activity of several neurons in a circuit and describe ensembles and attractor dynamics that provide data that are consistent with some neural network models. But, at the same time, we are still far from a rigorous demonstration of any neural network model with direct, yes or no validation experiments, as Popper argued should be done to prove or falsify scientific theories. So we still have not bridged the gap between elegant theory and data.

When will we know that we have bridged that gap? When will we understand how the brain works? Those are difficult questions to answer and depend on the particular viewpoint of what "understanding" means. I leave that to your own interpretation. In my view, a successful neural model should have quantitative accuracy to predict either the behavior and mental or perceptual state of the animal, or at least the future internal dynamics of the system. Richard Feynman famously argued that to gain a true understanding of a system, we must be able to actually build it, which is an even stricter definition of a successful theory than the one I just gave you. In a more humble definition, Kant insisted that knowledge is the systematic construct in which individual facts fit into a larger framework in order to make sense, like a house in which you can understand the role of bricks to know how the building would look. So, Kant argued, true knowledge occurs only when facts are connected in a ladder of explanation to other facts that are more complex (in this case, for example, the psychophysical and psychological states of a person) and are also connected to the lower rung of more basic facts (for example, the structural, molecular, and biophysical levels of neural circuits). So a proper paradigm for neuroscience would need to connect with the accumulated knowledge of these levels. Let's explore whether that is possible in the rest of the book.

▼ RECAP

Remember genetics, in which the genetic code is built with "codons," or triplets of nucleotides? A central question in neuroscience is, "what are the codons, or the foundational units of the 'neural code,' that relates neuronal activity with behavior or mental states?" If spikes are metaphorically the letters of the neural code, what are the words and grammar of the neural "language"?

The neuron doctrine argues that the unit of structure and function is the individual neuron, whereas neural network models argue that the brain is an emergent system and that groups of neurons, or ensembles, that have spatiotemporal correlations are the units of the neural language. This new paradigm may seem like a small change from the old one, but it has significant implications: it can not only explain spontaneous activity, but it also predicts that brains are intrinsically driven machines—machines that can learn and serve as universal computers. As we progress through the next lectures, we will examine how this paradigm can provide a common backbone that can help us put together a comprehensive view of what different parts of the brain do and how they do it.

Further Reading

Churchland, P. S., and T. Sejnowski. 1992. *The Computational Brain*. Cambridge, MA: MIT Press. A nicely written scholary review of early models of neural networks and their relevance to neuroscience.

Dayan, P., and L. F. Abbott. 2001. *Theoretical Neuroscience*. Cambridge, MA: MIT Press. A textbook coverage of computational neuroscience.

Hinton, G. E. 2000. "Computation by Neural Networks." *Nature Neuroscience* 3, suppl. 11: 1170. A review of convolutional neural networks by one of the pioneers.

Hopfield, J. J. 1982. "Neural Networks and Physical Systems with Emergent Collective Computational Abilities." *PNAS* 79: 2554–58. A mathematical description of the attractor model.

Hopfield, J. J., and D. W. Tank. 1986. "Computing with Neural Circuits: A Model." *Science* 233: 625–33. A provocative and accessible introduction to the emergent properties of neural network computation using the Hopfield model.

Seung, H. S., and R. Yuste. 2010. "Neural Networks." In *Principles of Neural Science*, ed. E. R. Kandel and T. J. Jessel. New York: McGraw-Hill. An introduction to neural networks, unfortunately omitted in later editions.

Trappenberg, T. 2009. *Fundamentals of Computational Neuroscience*, 2nd. ed. Oxford: Oxford University Press. A highly useful review of computational neuroscience.

Yuste, R. 2015. "From the Neuron Doctrine to Neural Networks." *Nature Reviews Neuroscience* 16: 487–97. A historical review of the conceptual paradigm switch in neuroscience, from the single neuron to a circuit-level framework of the brain.

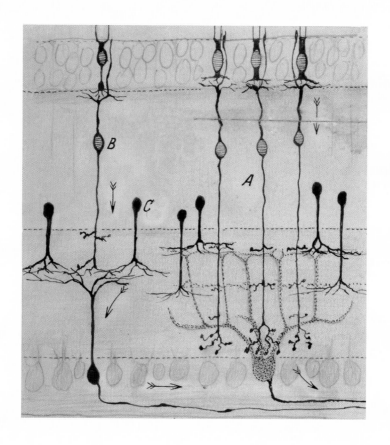

Neural pathways of the retina. The exquisite design of the retina, one of the better understood neural circuits, provides a beautiful example of the marriage of form and function in the brain. Courtesy of the Cajal Institute, "Cajal Legacy," Spanish National Research Council (CSIC), Madrid, Spain.

LECTURE 6: VISION

Things which we see are not by themselves what we see. . . . It remains completely unknown to us what the objects may be by themselves and apart from the receptivity of our senses. We know nothing but our manner of perceiving them.

—Immanuel Kant, *Critique of Pure Reason* (1781)

OVERVIEW

In this lecture, we'll discover

- ▶ How the retina works
- ▶ How object and motion information is processed separately by the cortex
- ▶ How color is constructed by the brain

At the core of the visual system is an act of interpretation, and for all intents and purposes, what we think of as reality is a simplified rendering of the world that for hundreds of thousands of years has enabled humans to interact with our environment and other humans, stay alive, and reproduce. But what we see may not be exactly what is actually there. As Plato and Kant argued, what we see is but a glimpse of the real world. Going back to our discussion in the first lecture, what we perceive is likely a constructed model of the world, which we use to predict the future. Perhaps nowhere is this more evident than in our visual system.

Vision is the crowning achievement of sensory evolution. Even Darwin famously said in the middle of *On the Origin of Species* that he felt inadequate trying to explain how evolution could ever have come up with something as sophisticated as the design of the human eye. This is what he wrote:

> To suppose that the eye with all its inimitable contrivances for adjusting the focus to different distances, for admitting different amounts of light, and for the correction of spherical and chromatic aberration, could have been formed by natural selection, seems, I confess, absurd in the highest degree. . . . The difficulty of believing that a perfect and complex eye could be formed by natural selection, though insuperable by our imagination, should not be considered subversive of the theory. (Darwin, 1872)

Vision is magnificent, and as primates, our visual system is arguably our most important sense. It dominates the brain in terms of size—occupying a significant portion of our cortex—and also in terms of the amount of information it brings to the brain, probably more than all the other senses combined. We use vision for many purposes: we determine the location, shape, and size of an object, as well as its movement, direction, speed, and also its color, which gives us information about its surface. We can do all this at a distance, and if you have good eyesight, even at a considerable distance. If that weren't enough, vision is sensitive over ten orders of magnitude in light level. Yes, this is not a typo: a 1 followed by 10 zeros! Your eyes can even detect single photons, which is the smallest possible unit of light. No wonder vision dominates the mammalian sensory system.

We usually assume that vision shows us the world the way it is. We are wrong. As we'll see in this lecture when we dig deeper into the visual system, we find that it operates by carefully building up images and assembling them into a coherent structure. As Kant argued, the visual system constructs the world that we see in its essence, extracting particular bits of information and combining them. It doesn't make up the world but interprets it as it sees fit. Why does it do this? Our visual system is designed to allow us to perceive the world in a way that maximizes our chances for survival by building predictive models. The purpose is not to accurately record what is there but to help us survive.

Let's look at how this is done. But first, we need to take a survey of our visual system and touch on three basic principles of neuroscience that we have already encountered: hierarchies, maps, and parallel wiring.

Anatomy

The visual system is organized as a pathway (figure 6.1). As mentioned, it starts with the retina at the back of the eye, then progresses to the lateral geniculate nucleus (LGN) of the thalamus and continues to the cortex, first to the primary visual cortex area and then to secondary visual areas

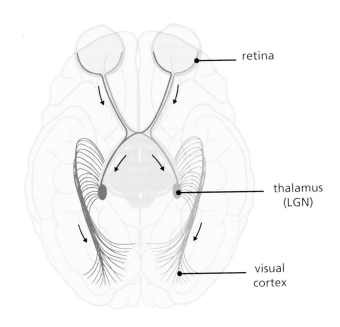

6.1 The **visual pathway**, viewed here from the bottom of the brain, progresses from the retina to the thalamus to the cortex, building topographic maps of the world at each stage.

that spread out through the dorsal and ventral regions of the occipital lobes of our brains.

The visual pathway illustrates three principles. First, as we learned from the Hubel and Wiesel model, this pathway is *hierarchical*, meaning that as you move to the next step, neurons do something more sophisticated, i.e., have more complex receptive fields than the neurons before them, forming a hierarchy of processing units.

Second, the visual system is *organized topographically* in maps. Each step of the pathway forms a map of the visual world. That means that objects that are next to each other in the world are detected by neurons that are next to each other in the brain, so the visual system creates a faithful representation of the physical structure of the outside world. We have maps of the world in our heads. These maps can be distorted, but they are still topographically and beautifully organized.

Third, there are many *parallel pathways*, not just one. Together they ascend from the retina into the brain, side by side Each of these visual streams processes a particular type of information and ends at a different place in the brain. We are still not sure how many pathways there are. But these parallel pathways are not completely isolated from each other: they are called *streams*, and you can think of them as many streams flowing through valleys and plains, sometimes joining and separating again, but they're all part of the same river.

Retinal Phototransduction

Human eyes are spherical in shape, which allows them to move smoothly. They invert the light through the pupil and focus it, using the cornea and the lens, onto the *retina*, where active processing of light information begins (figure 6.2). The retina is a thin layer of neural tissue at the back of the eye that, along with other types of cells, contains photoreceptors. Each retina has about 10^8 photoreceptors, each taking in photons of light and turning them into electrical signals.

Interestingly, photoreceptors are located at the back of the retina, the farthest from the light, which has to cross the entire retina to reach them.

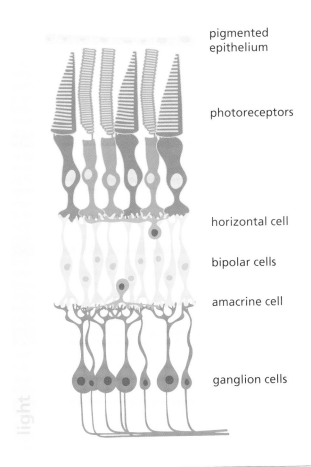

pigmented epithelium

photoreceptors

horizontal cell

bipolar cells

amacrine cell

ganglion cells

light

6.2 The **retina** is a layered neural circuit with five basic cell types, and is built to capture photons efficiently and transform them into electrical signals.

This might seem counterintuitive, but a likely reason is that behind the retina, all the way in the back, is the *retinal pigment epithelium* (RPE), which provides support and nourishment to the photoreceptors. The RPE supplies photoreceptors with photopigment and, in addition, it prevents scattering of light and vision distortion by absorbing the stray photons that made it past the retina. Without the RPE, photons would bounce around inside the eye and degrade image quality.

There are two types of photoreceptors: rods and cones (figure 6.3). Rods are extremely sensitive to light, and we rely on them in low-light conditions. In contrast, cones are adapted for detecting bright light. Rods and cones have a similar structure, with an outer segment that contains

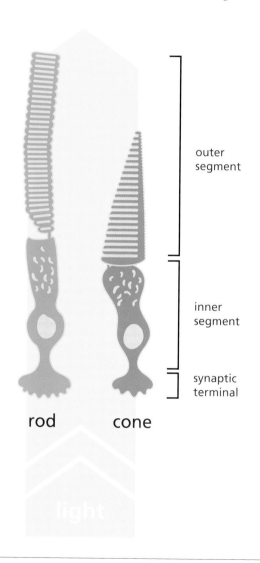

outer segment

inner segment

synaptic terminal

rod   cone

light

6.3 The two types of **photoreceptors**, cones and rods, have peculiar shapes that are nicely adapted to their function stacks of discs with very high densities of the visual pigment (called *rhodopsin* in rods and *conopsin* in cones). Photoreceptors also have an inner segment, which contains the Golgi apparatus, the endoplasmic reticulum, and a lot of mitochondria, as well as a synaptic terminal, where the neurotransmitter glutamate is released, activating the rest of the retinal circuitry. Photoreceptors are elongated in the direction of the light and likely function as tiny optical guides that funnel photons efficiently into the discs in the outer segments. In fact, the outer segment discs are precisely stacked perpendicular to the light to ensure that every photon that passes by can be absorbed, maximizing the chance that it will encounter a photopigment. Photoreceptor cells are so efficient that they enable us to detect extremely low numbers of photons, close to individual ones. As mentioned, this design is optimized to the physical limit: a single photon is the elementary unit of light: you cannot do better than that!

The reason photoreceptors are so good at detecting photons lies in the outer segment discs, where magic happens: photons are turned into electrical currents. This process, "phototransduction" (figure 6.4), is relatively well understood and is also quite peculiar: the absorption of a photon leads to the hyperpolarization of the cell and the shutting off of its synapses. The relationship between light and transmitter release is thus inverted: in the dark, the photoreceptor is *depolarized* and the neurotransmitter glutamate is released, whereas in the presence of light, the photoreceptor *hyperpolarizes* and transmitter release ceases. In other words, photoreceptors are turned off by light! Weird!

Let's dive into this. In the dark, photoreceptor cells have a lot of cyclic guanosine monophosphate (cGMP), a cyclic nucleotide derived from guanosine triphosphate (GTP), which opens cGMP-gated ion channels, which in turn lets sodium into the cell and depolarizes it. But when a photon hits the outer segment, it becomes hyperpolarized by a series of biochemical reactions that cause the levels of cGMP to drop. This process seems awkward, but very interesting: the photon is absorbed by transretinal, the photopigment in the membrane of the stacks of the discs in the photoreceptor outer segments. That absorption bends the retinal structure, activating the rhodopsin, which holds it in a pocket. Activated

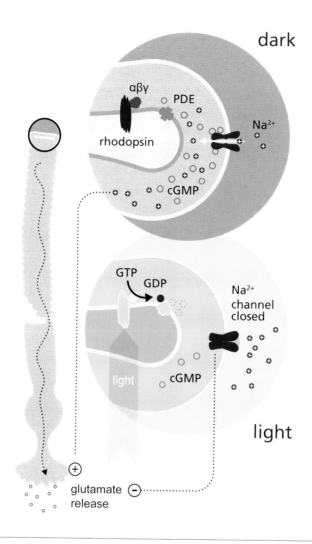

6.4 Phototransduction occurs in the outer segments of the photoreceptors, when incoming photons are absorbed by rhodopsin, activating a biochemical pathway that closes sodium channels, hyperpolarizes the cell, and turns off synaptic transmission.

rhodopsin then activates transducin, a G-protein in the membrane of the disc, which diffuses and activates an enzyme—phosphodiesterase (PDE)—which in turn quickly chews up cGMP, reducing its concentration in the cytoplasm, and that closes the channels and hyperpolarizes the cell.

Why so many steps? For amplification! Each time an activated molecule turns on several other downstream molecules, that essentially performs a multiplication. So the absorption of a single photon ends up causing a significant change in the current fluxed by the channels. But the problem is, the more steps, the slower phototransduction occurs. Vision is in fact a slow sense; it can take 30 ms for the retina to react to a visual stimulus, which is a long time compared with, say, the auditory system. Do we really want vision to be slow? Maybe not, but this slowness could be the price we pay for sensitivity. You cannot have it both ways. But there is something we can do to speed things up. You may know from physics that diffusion is much faster in two dimensions than in three, as Einstein showed. So, by constraining the biochemical machinery of phototransduction to the membrane of the stacks, which is two-dimensional, as opposed to the cytoplasm, which is three-dimensional, nature is speeding the reactions tremendously. Excellent, now it all makes sense. Well, almost. So why does light turn the photoreceptor off instead of on? Honestly, I don't know. I have read many hypotheses but am not convinced by any of them. Maybe one of you can help explain why nature chose to build phototransduction backward, as if we are detecting darkness and shades rather than light and objects.

Retinal Processing

Now that we understand what photoreceptors do, let's turn our attention to the rest of the retina. In addition to photoreceptors, there are four other types of cells in the retina: horizontal, amacrine, bipolar, and ganglion cells (figure 6.2). Photoreceptors reciprocally connect to horizontal and bipolar cells, and bipolar cells connect to amacrine cells, which then connect to ganglion cells. Ganglion cells are the clinchers: they are the only retinal cells that actually fire action potentials and that have axons that leave the retina, so they generate all the retinal output in the form of electrical signals. To understand what all the retinal circuitry is doing, let's skip the intermediate stages and jump to the ganglion cells, since they are the final step in the retinal processing of light. We will then return to see what neurons in the intermediate steps do.

Understanding of the visual system is closely associated with the understanding of visual receptive fields, the discovery of which was made possible by single-cell (i.e., single-unit) recordings, as I mentioned in earlier lectures. In case you forgot, a fine-tipped electrode is inserted into the tissue and carefully placed either within the cell membrane of a target neuron or just outside it, to allow for intracellular and extracellular recordings. The activation or inactivation of the neuron is then measured by an increase or decrease in the frequency of the action potentials, and the stimulus that activates the neuron is called the *receptive field*. So to understand what a neuron does, if you map its receptive field—that is, the spatial or temporal properties of the stimulus that makes it fire—you are done, at least according to the neuron doctrine.

Although researchers had recorded retinal receptive fields before, our story starts in the 1950s. Stephen Kuffler was using microelectrodes to record action potentials from retinal ganglion cells of cats. He used a modified ophthalmoscope to study responses of the retina to a diffused background light and a highly focused stimulus spot. A small bright spot was shone on different parts of the retina and moved around to map the receptive field of the neuron (figure 6.5).

Kuffler found that ganglion cells had antagonistic and concentric center-surround receptive fields. In some cells, the "ON-center" cells, shining the light in the center of the receptive field produced strong responses, but that response was inhibited when the spot of light increased, encompassing what we call the *surround* of the receptive field of the cell (figure 6.5). Interestingly, when he illuminated only the surround, the ON-center ganglion cells were completely shut off. Meanwhile, in other cells, the "OFF-center" ganglion cells, the neuron was inhibited if light was shone on the center of the cell but activated when there was light in the surround.

In other words, retinal ganglion cells have on/off center-surround receptive fields (figure 6.6). Because of their antagonistic surrounds, ganglion cells respond more vigorously to small spots of light rather than large spots of uniform illumination of the visual field. Now, note that we are talking about essentially all ganglion cells: they respond not to diffuse light but to changes in light intensity between their center and

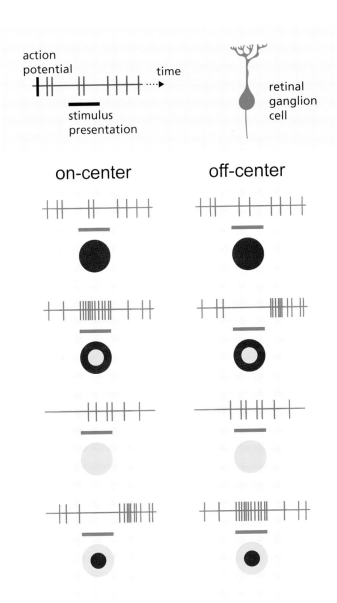

6.5 Retinal ganglion cells have on-off center-surround antagonistic receptive fields, responding to the presence or absence of light in the center or surround of their receptive fields.

ON-center   OFF-center

6.6 On/Off receptive fields can be represented with pluses or minuses to denote excitation or inhibition of the center-surround territories.

surrounds. Fascinating. It seems that the retina is interested in measuring differences in light intensity (luminance) in the borders of objects: it is those that, because of their sharp light/dark changes, stimulate the retina the most. We refer to that as "contrast": the relative difference between stimuli, which you can define mathematically as (stimulus 1 − stimulus 2)/ (sum of stimuli 1 + 2).

Now, let me point out that contrast is a calculation performed by our neurons on the image of the stimuli: it's a function that we can compute, about the world. This is an important point, and not just about vision. Most of our sensory systems don't care about absolute intensities of stimuli; rather, they take note of relative differences in intensity, computing stimulus contrast. Why? Let's recall the ideas from the first lecture. The brain is interested in measuring changes, because that's the key information it uses to update its predictive models of the world—models that incorporate how things change. Who cares about the brightness of an object? That's boring. What's interesting is how the object changes in space and time. So it seems that from the very beginning in the retina, the visual system is interested in changes, predicting new things, the saliency of the stimulus. And by using center/surround receptive fields, it obtains the most information about object boundaries.

Now you are probably wondering why retinal ganglion cells come in two types—on and off—whereas I emphasized that all photoreceptors are turned off by light. Then shouldn't all the retinal ganglion cells be turned off by light as well? The reason that about half the retinal ganglion cells are turned on by light involves the retinal circuitry. Photoreceptors connect to bipolar cells, and some bipolar cells have special glutamate receptors that hyperpolarize the cell rather than depolarize it. They thus flip the sign of the electrical signal from the photoreceptors so they get turned ON rather than OFF by light. These ON bipolar cells then excite ON ganglion cells. The rest of the bipolar cells are better behaved, just relaying to the retinal ganglion cells whatever the photoreceptors are doing; that's why they are called OFF *bipolar*.

Now that I've explained how ON and OFF responses are built by the retina, you may be curious how the center-surround responses are generated. That's where some of the other retinal cells—the horizontal cells—come to the rescue. They are reciprocally connected to the photoreceptors and are excited by them, but they in return inhibit both photoreceptors and bipolar cells. That means they flip the sign of whatever signals they receive, so a depolarizing (excitatory) signal from a photoreceptor becomes hyperpolarizing (inhibitory), and vice versa.

These horizontal cells are connected in the following way. They receive inputs from many neighboring photoreceptors and inhibit all of them back, creating a blanket of inhibition that surrounds the area that was turned on, or, conversely, a blanket of excitation (technically, *disinhibition*) that surrounds the area that was turned off. That's why you find center and surround regions of the receptive fields with opposite polarity. This works regardless of whether the center is OFF or ON, as the horizontal cells do the opposite of what the photoreceptors do.

Retinal Functions

By computing spatial contrast, the retina is turning the visual scene into an outline of the borders of the objects, which is tremendously useful if you want to recognize objects, as we will see when we talk about the visual cortex. But the retina also does many other things. There are probably more than forty types of ganglion cells in mammals and probably as many functions of the retina as types of retinal ganglion cells. The idea is that each of the retinal ganglion cell types starts a pathway that propagates through the brain. So you can think of the retina as the beginning of a set of parallel pathways of visual processing.

Among other things, the retina processes information about changes in lightness (ON pathway) versus changes in darkness (OFF pathway), since it has retinal ganglion cells that respond to increases or decreases in stimulus luminance (in stimulus contrast, to be more precise). The retina also has specific pathways to process visual information during the day (the pathway that starts with cones) versus night (the rod pathway). Additionally, some retinal ganglion cells have larger or smaller receptive fields, and specialize in detecting larger or smaller objects. A similar thing happens with temporal responses: some retinal ganglion cells have brief, rapid responses, as opposed to others that have slow and sustained responses. Also, retinal ganglion cells compute temporal contrast. All that could come in handy if you want to track objects that are changing or moving quickly versus slow or stationary objects.

There are also other types of retinal ganglion cells in some species that are specifically designed to detect movement or cancel movement of the eyes or the head. This fact echoes the overall theme that to build predictive models of the world, the brain is not interested in static things but in dynamic changes in space and time. There are also ganglion cells that can detect color, and we'll learn more about them later in the lecture. Finally, a few ganglion cells contain photopigment, acting as photoreceptors, and care about the absolute luminance of the visual scene instead of contrast; their output is used to reset and control our circadian clock.

And that's not all. It turns out that all retinal ganglion cells adapt to changes in visual stimulation, meaning that their responses change when you change the properties of the stimulus—for example, its light level. Some people say that light adaptation is the main reason we have a retina. You may have noticed how professional photographers constantly measure luminance of the scene so their picture is not too dark or too bright? The retina does that for us automatically. A detector that is good at adapting to light levels is exactly what you need if you live on Planet Earth, where there are massive (I mean by eight to ten orders of magnitude) changes in luminance across the day-night cycle and also within the course of a day, depending on the weather. If extraterrestrials were to examine our retina, they would conclude that it was designed for living in the exact light conditions of our planet.

A final note before we leave the retina: I have described only a partial list of functions of the retina. Even though it's always argued that the retina is the best-understood part of the brain, we still don't know the functions of most subtypes of retina ganglion cells.

From Retina to Cortex

So now let's leave the retina and climb our way into the brain. The visual pathway is pretty straightforward: retinal ganglion cells project to the lateral geniculate nucleus, where they make connections with neurons that project to the primary visual cortex. Thus, with one single stop in the thalamus, you jump from the retina to the cortex (figure 6.7). There are other side pathways from the retina that need to be mentioned. Some retinal ganglion cells project to the superior colliculus, which controls reflexes and the orientation of the movement of the head and eyes. And

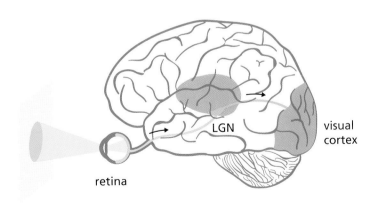

6.7 The **visual pathway** reaches the cortex after a stop in the lateral geniculate nucleus (LGN) of the thalamus.

then the aforementioned luminance-measuring retinal ganglion cells project to the suprachiasmatic nucleus (SCN), which controls our daily (circadian) rhythms.

But the bulk of the action happens in the main visual pathway, where the vast majority of retinal ganglion cells project through the optic nerve to the LGN in the thalamus—the thalamus being the main relay station of the senses, if you will. Now, picture this: each optic nerve extends from each eye, but instead of continuing straight to the brain, there is a crossing over to the other side at a point called the *optic chiasm*. At that point, ganglion cells' axons from the medial (nasal) half of the retina project to the other side of the brain, while lateral (temporal) retinal axons stay on the same side. So at the chiasm, axons from one *temporal* hemiretina join the axons from the contralateral *nasal* hemiretina to form the optic tract. From there, axons go to the LGN together, where ganglion cell axons make contact with thalamic neurons. These neurons send axons to the primary visual cortex via the *optic radiation*.

This crossing over of axons from medial hemiretinas is not accidental. Because the eye inverts the image, this crossover makes sure that all the information arising from the left side of the visual field ends up in the right side of the brain, and vice versa. Peculiar, no? The crossing over of sensory axons is a general feature in the brain, as we will see. The evolutionary origins and purpose of this crossing over to the contralateral side are shrouded in mystery. One idea is that because the eyes invert the image and given that vision is a dominant sense, vision crosses first and the rest of the sensory systems follow vision. But why would vision need to cross? I have read many theories but remain unconvinced, as I can also imagine that nature could have kept all the information uncrossed, since it doesn't really matter where it is processed. Anyway, as the Royal Society motto says, quoting Horatio, *Nullius in verba* ("take nobody's word for it").

Let's head up into the LGN. What is it doing? Well, we still don't quite understand its function. That is yet another neuroscience mystery waiting for someone like you to solve. Receptive fields in the LGN are similar to those of retinal ganglion cells. Researchers have traditionally argued that the LGN is just a "relay" nucleus. But there are actually ten times more LGN neurons than retinal ganglion cells. Surely all these LGN neurons must be doing something more than just passing the ball upstairs, otherwise you could simply run the axons from the retinal ganglion cells to the cortex and save yourself the trouble of building and maintaining the LGN, no? (By the way, this mystery of the function of the thalamus applies to all sensory modalities, not just vision.)

LGN connectivity perhaps provides a hint to its function. The connections from the retina to the LGN pathway are all *afferent*, meaning they go *into* the brain. There are no *efferent* connections (out of the brain) from the LGN back to the retina. But when it comes to connectivity from the LGN to the cortex, for every afferent connection going up to the cortex, there are ten efferent connections that come back to the LGN. That's also the case for all thalamic nuclei, which suggests that the thalamus is really working for the cortex; that is, it is mainly driven not by the sensory periphery but by the cortex. Why would it do that? Perhaps from all the information that is coming into the thalamus from the senses, the cortex is selecting and turning on certain neurons that it is interested in "listening" to, as a selective filter. That could be the basis of attention: the cortex could be selecting which inputs it attends to, and this could happen at the thalamus.

The strong cortical modulation of sensory inputs by the cortex is also consistent with the argument that the brain constructs the world. A step further in this direction is the possibility that the thalamus builds not just our attention but our awareness and consciousness as well. Indeed, widespread lesions in the thalamus render patients unconscious. So the thalamus could be doing something quite exciting: building our sense of self. I know you want to hear more about this, but you'll have to wait until the last lecture, where I discuss consciousness.

Here is a final comment on the LGN, going back to the principles of neural development. Remember when we discussed whether or not the brain is wired in a specific way? The connections from the retina to the LGN are one of the most striking examples of connection specificity, as, in primates, each LGN neuron is innervated by one retinal axon, and both retinal and geniculate cells have similar receptive fields. LGN cells are arranged in layers, and different layers are innervated by different eyes, with ordered retinotopic maps of the visual field that are aligned one on top of another. And this precise structure assembles itself. Wow! How is that for specific wiring! When you look at this wiring job, you could imagine that the entire brain could be wired with utmost precision. The basis of this wiring could be preprogrammed genetically, but activity-dependent pruning could also be at play. In the fight for the "soul" of the brain, whether it is precisely or randomly wired, the LGN falls squarely in the precision camp. But let's move up to the cortex before we consider that this battle is over.

Visual Cortex

After Kuffler discovered the center-surround receptive fields, he was directing the work of two rebellious students in his lab at Johns Hopkins, David Hubel and Torsten Wiesel (my scientific family and now like family to you too) They were following their mentor's research performing single-cell recordings in the retinas of cats, and they asked the natural follow-up question of what happens to the retinal signals as they go up to the LGN. But when Hubel and Wiesel recorded receptive fields of LGN neurons, they saw the same ON/OFF center-surround receptive fields that Kuffler had found in the retina. So they got bored and moved up to the visual cortex.

The experiment was this: they placed an electrode in the middle of the visual cortex of an anesthetized cat to record the activity of individual neurons (figure 6.8). Then, they tried to stimulate the animal's visual field by showing a small dark circle of metal glued to a glass microscope slide, the same stimulus that was highly effective at eliciting a response from the retinal ganglion cells and the cells in the LGN. But they got stuck—they couldn't get cortical neurons to fire! Day after day, week after week . . . nothing. Then, one day, a lucky break. Hubel and Wiesel noticed that a neuron started firing when they were removing the slide from the projector. Repeating the same movement over and over again, they finally realized that the cell was responding to the shadow that the slide cast when being inserted. What serendipity! This lucky break enabled them

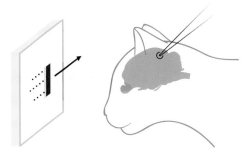

6.8 Hubel and Wiesel discovered, by chance, that most neurons in the primary visual cortex respond to **oriented bars**.

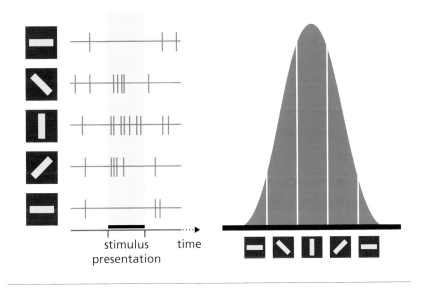

stimulus   time
presentation

6.9 Visual cortex neurons have **orientation selectivity**, i.e., they respond to elongated visual stimuli with a particular orientation.

to discover that neurons in the visual cortex respond to edges, to bars of lights, or bars of shade (figure 6.9). As mentioned in lecture 4, visual cortex neurons have receptive fields that are elongated (they are no longer concentric). Let's revisit this now, but in more depth.

Figure 6.9 represents single-cell recordings from the primary visual cortex, VI. The neuron does not respond to a horizontal bar, it fires above the resting level when the stimulus is placed diagonally, and it fires the most when the stimulus is vertical in orientation (figure 6.9). The conclusion is that VI neurons can distinguish between the orientation of objects in the visual field, an ability that we call *orientation selectivity*. How do the receptive fields change from being concentric to bar-shaped? That's why Hubel and Wiesel proposed their famous model, as discussed in lecture 4. The assumption was that the shape of the receptive fields in VI is "constructed" from inputs of several LGN neurons added together (figure 6.10). So the oriented receptive field was the consequence of adding together a bunch of circles all lined up.

In fact, Hubel and Wiesel observed many different receptive fields—neurons responding to different orientations of the bar, different widths, and different center-surround relationships. They called them all *simple cells*;

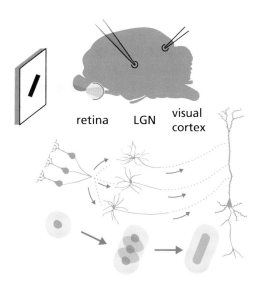

retina   LGN   visual cortex

6.10 According to **Hubel and Wiesel**, simple cells' receptive fields are built when LGN cells with concentric receptive fields synapse on a cell in V1. As a result, the cell in V1 gets an expanded receptive field that is a sum of the three separate receptive fields.

we have talked about this already, but repeating important information in class makes it stick. Simple cells respond only to a bar of a particular orientation in a particular part of the visual field. But, as I mentioned in the last lecture, they also found cells with more complicated receptive fields, which they correspondingly called *complex cells*. These neurons responded to the orientation of bars positioned anywhere within the visual field of the animal (maybe not all over the visual field but in a much larger part of it). To explain complex cells, Hubel and Wiesel extended their simple-cell model and proposed that complex cells could be generating larger receptive fields by summing up inputs of different simple cells (figure 6.11). This relationship builds a natural hierarchy of processing, with the cells that are "higher up" in the chain knowing more about the world than the ones reporting to them. Simple cells allow you to see an object only if that object is positioned in a particular part of the visual field, while complex cells can detect that object in multiple locations—which, by the way, suggests that complex cells are probably involved in tracking movement.

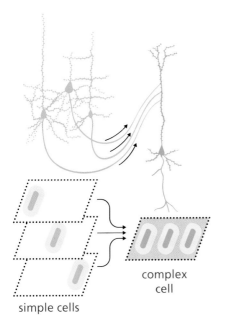

simple cells

complex cell

6.11 Complex receptive fields are built when several simple cells project to a complex cell. This enables the complex cell to respond to oriented bars in a large area of the visual field.

As explained in lecture 4, this hierarchy continues higher and higher until you reach neurons that fire as a response to specific objects. The classic example is the grandmother cell—a neuron in the secondary visual cortex that fires only when you see your grandmother. This way, the visual system, as a good feedforward neural network, has taken the problem of recognizing objects and spread it out in a hierarchy of processing in which each layer is doing one job. And as you go up, you are abstracting from the world, all the way from the contrast computations in the retina to recognizing your grandmother (figure 6.12). As discussed in the first lecture, hierarchy of information processing is one of the principles of neuroscience, and this model of vision can be described as a perceptron.

But Hubel and Wiesel didn't stop there. In addition to discovering simple and complex receptive fields, they found that the primary visual cortex is organized in columns. If you push an electrode straight down from the

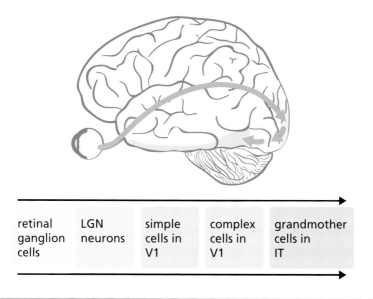

| retinal ganglion cells | LGN neurons | simple cells in V1 | complex cells in V1 | grandmother cells in IT |

6.12 The visual system builds increasingly more sophisticated receptive fields in a **hierarchy** of processing steps, from center-surround retinal receptive fields to the recognition of faces in inferotemporal cortex.

surface of the cortex to the bottom and record from the different cortical layers in the same "column," you would see that the neurons have similar receptive fields and similar orientations, as if they were forming a functional unit. But, look at what happened: they inserted the electrode at a flat angle and moved it horizontally and found that neurons they encountered now had different orientations. So cortical columns are arranged to form a map of orientations, and by moving horizontally, you shift from one orientation to the other systematically. Think of it: we have a map of the orientation of the visual stimuli imprinted in our primary visual cortex.

What does this orientation map look like? Instead of using electrodes to measure it, you now also measure it via optical imaging methods. You can take a picture of the surface of the cortex from above and look at the reflectance of hemoglobin optically. The spectroscopy of hemoglobin is related to oxygen consumption, which is itself related to neuronal activity. Using this functional imaging method, if you were to assign different colors to neurons according to the different orientations of the bar they respond to (as Amiram Grinvald and Tobias Bonhoeffer, two researchers from the Wiesel lab, did while I was doing my thesis there), you would find beautiful maps that resemble pinwheels. These pinwheels are a structured representation of orientation selectivity territories organized around nonoriented centers, with each orientation arching from it, like the blades of a pinwheel (figure 6.13). Isn't nature spectacular? Why pinwheels? They could be an elegant solution to the computational problem of how to map things. Developmentally programmed? Efficient use of space? More mysteries.

To make it more complicated and beautiful, it turns out that there is actually more than one functional map in the visual cortex, and that all these maps are superimposed on top of one another. Besides the pinwheel map of orientation selectivity, we have an ocular dominance column map, which you should remember from lecture 2, dividing the visual cortex into regions that preferentially respond to one eye or the other. In addition, there are regions in the middle of the ocular dominance column (called *blobs*) that respond preferentially to color. You can explain all these maps with the hypothesis that the visual cortex is built of modules (figure 6.14). A module (known as a "*hypercolumn*") is a part of the

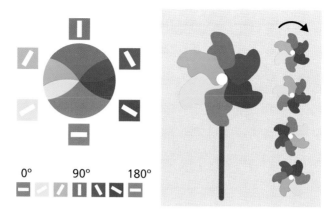

6.13 The territories of cortical neurons that respond to particular stimulus orientation are organized in **"pinwheels,"** areas where orientations are mapped systematically around a center of non-oriented cells.

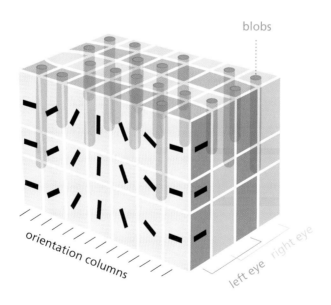

6.14 Cortical modules are units of visual processing that contain superimposed maps of orientations, ocular dominance, and color.

cortex that "houses" all the neurons that respond to a particular area of the visual field. This hypercolumn would have two ocular dominance columns (one from each eye), each with pinwheels of orientation, and blobs that carry information about color. And there are other variables mapped in V1, one being spatial frequency of the stimulus, and likely many others.

Thus, in the visual cortex, we find key principles of neuroscience demonstrated: hierarchical processing, modular organization, and topographic maps. What is the function of all these maps? The primary visual cortex decomposes the image into a series of what are known as *visual primitives*. The word "primitive" is used in a mathematical sense—primitives of a function refers to its essential building blocks Examples of primitives that the primary visual cortex is computing from the world include retinotopy (location), edges, depth (distance from the observer), colors, motion, and shapes. Combining these primitives as variables in a set of mathematical equations, you could perform any visual computation. But to explore that, we need to leave the primary visual cortex and move higher up in the hierarchy.

Dorsal and Ventral Pathways

There are two main objectives of the visual system: recognizing objects and tracking their motion—both logical needs, as we not only have to recognize the prey (or the predator) but also run after (or from) it. The same argument applies to recognizing running to (or away from) a potential mate, as reproduction is right there with survival as a critical evolutionary need.

How does the brain do this? Once visual information has been processed by V1, it moves on to higher cortical areas in two separate pathways, known colloquially as the *what* and *where* pathways. The what, or *ventral*, pathway runs downward toward the temporal lobe (figure 6.15). The where, or *dorsal*, pathway moves from V1 to the parietal lobe brain They split the job: the what stream is devoted to processing shapes and the where pathway processes motion.

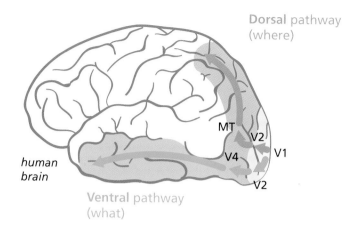

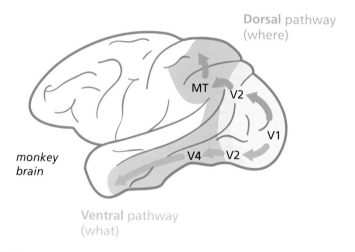

6.15 After the primary visual cortex, visual information spreads through the cortical areas, following the **"what"** and **"where"** pathways, specialized in computing object shape or motion.

The Where Pathway

Let's first take a look at the dorsal pathway, which, incidentally, begins in the retina. Motion computation starts in a group of particularly large ganglion cells in the retina, the "magnocellular" (M for *magno* = large) cells. M cells have larger receptive fields, which means they don't focus on details of objects but, rather, on larger areas. They also respond rapidly. That makes sense, as motion means change of position over time ($\Delta x / \Delta t$), so you need to measure a large space as quickly as possible to make accurate motion calculations. M retinal ganglion cells project to two so-called M layers of the LGN (layers 1 and 2, one for each eye). Axons from these neurons go to layer 4 of V1, and from V1 the pathway then ends up in the secondary visual cortex (V2), and upward, mostly branching into the parietal lobe.

To understand what happens with motion information in a Hubel and Wiesel fashion, let's insert electrodes and record receptive fields in those higher cortical areas. What we find is that neurons have large and rapid-responding receptive fields, consistent with those of M cells but which are now also sensitive to motion of the object. There's an area called *MT*, for *mediotemporal*, where most neurons are selective to the direction of motion of objects. Neurons can fire if there is slight motion in the image. Moreover, if patients have a lesion in this area, they become "blind" to movement, meaning that they cannot detect the movement of correlated dots, at levels at which people without such a lesion have no difficulties (figure 6.16). This syndrome is called *akinetopsia*, or movement blindness;

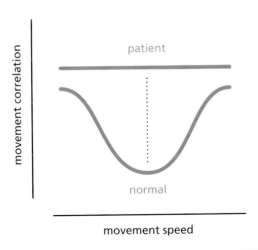

6.16 Lesions in **visual area MT** prevent patients from detecting movement, as shown by the higher correlation needed to detect dots moving at different speeds.

and patients with this condition could be in danger when they cross busy New York streets, as they also have difficulty detecting moving cars.

The What Pathway

Let's now look at the other pathway, the ventral (what) pathway. It also starts all the way back in the retina in a series of small retinal ganglion cells, the *parvocellular* (or P, for *parvo* = "small") cells. They have smaller dendrites, so their receptive fields are smaller, which is useful for mapping the world with precision. They are also slow to respond, since it pays to measure things slowly if you want to determine their position. P retinal ganglion cells feed into parvo layers of the LGN (layers 3–6). These LGN parvo cells project to particular layers of V1 (different from the layers of the magno pathway), with both simple and complex receptive fields. From there the P pathway projects to V2 (to different regions than the M pathway) and from there to higher cortical areas, mostly in the temporal lobe.

What happens if you insert an electrode into cortical areas from the ventral pathway? Let's say you record from a neuron in the inferior temporal area (IT). You find neurons that respond specifically to faces (figure 6.17). In fact, they respond even better to sketches or cartoons of

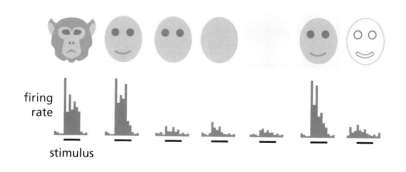

6.17 Consistent with the grandmother cell idea, neurons in the **inferotemporal** cortex of monkeys and humans selectively respond to faces.

faces than faces directly. That may be why we like cartoons, as they represent the pure essence of the object.

Face-selective neurons are widespread in the infratemporal cortex of monkeys and humans, and there are reports of electrode recordings from epilepsy patients with neurons that are selectively activated by the faces of Jennifer Aniston and Bill Clinton. If you know who they are, you must have similar face cells too!

People with lesions in their ventral pathway experience cognitive deficits called *agnosias*, from the Greek a-gnosis, "not knowing." These are fascinating syndromes. A famous one is prosopagnosia, in which the patient cannot recognize faces. They see the person and can describe the face but cannot identify who it is. A wonderful book from the late Oliver Sacks, *The Man Who Mistook His Wife for a Hat*, tells the story of one such patient, who couldn't recognize the face of his wife even though he could see her perfectly well and recognize her voice.

Another example of ventral stream lesions are "vegetable" agnosias, where people cannot recognize vegetables or a type of vegetable. They can see them perfectly well and describe their color and shape, but they just don't know what they are. They are gone from their mind! Vegetable agnosias suggest that neurons that respond to vegetables have been damaged, and since brain lesions tend to be localized, this means that most neurons responding to vegetables are likely in proximity to one another. So neurons that recognize particular objects in the world must be arranged in maps in our temporal lobe. Given that agnosias tend to be selective for categories of objects, these maps in the brain must follow some sort of order. It looks like we have a map of the world in our heads!

Color Vision

Now that we understand how people recognize objects and their movement, let's talk about how colors are processed in the visual system. In fact, the way color perception is generated in the brain is particularly insightful as it provides a glimpse deep into the inner workings of our mind. The punch line is that color is constructed, so that could make Kant particularly happy.

The story begins in the photoreceptor cells in the retina with funny-looking hats, the cones (see figure 6.3). There are three types of cones, which have three different opsins tuned to absorb light of different wavelengths: short, middle, and long (figure 6.18). We know that cones are important for color perception because humans who have colorblindness often have two or even one cone opsins as opposed to three. But it would be incorrect to say that individual cones cells "see" color: in fact they can't. They respond exactly the same way to a few photons of a sensitive wavelength as to many photons of a less sensitive wavelength.

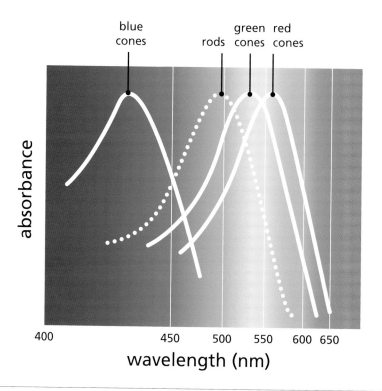

6.18 The **visual spectrum** is measured by our retinas using differential absorption of opsins in four types of photoreceptors.

What, then, creates color? Our brain compares the responses of different cones to light and calculate which color is seen. This process begins with specific parvo retinal ganglion cells, which receive inputs from cones. Their receptive fields are not simply ON or OFF center, but they also respond to a specific wavelength combination, such that some neurons respond preferentially to green or red in the center. The surround is always the opposite color (*color-opponent* cells). The color combinations are of two types: green-red or blue-yellow (blue and yellow are generated by mixtures of inputs from green and red cones). These *color-opponent* cells (figure 6.19) help the brain calculate color by computing the color contrast—again, defined as the normalized difference between a receptive field center and its surround.

From these special color-opponent retinal ganglion cells, the color pathway continues up to P LGN layers and then to round structures called blobs

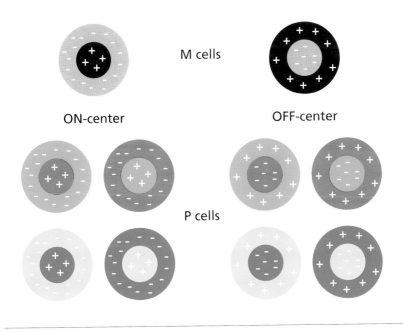

M cells

ON-center   OFF-center

P cells

6.19 As opposed to rods, which are wavelength insensitive, **color-opponent** retinal ganglion cells respond to combination of wavelengths in the center and surround of their receptive fields.

in the middle layers of V1; from there the pathway goes to the temporal lobe together with the what pathway. There we find a cortical region, named V4, that is full of cells that respond to color; and by "color" I don't mean wavelength but, rather, our perception of color, which are two different things. This is important. It turns out that the wavelength emitted (by reflectance) by an object depends on the wavelength that hits it. The problem is that, in nature, the light that reaches the earth changes in wavelength throughout the day and also depends on cloud coverage. Thus, the wavelength reflected from objects also changes during the day. Because of this, just measuring wavelengths would not be a very useful way to identify objects, since, for example, a tiger that looks yellow at midday may appear reddish at dusk. That's where the brain comes in: by computing the contrast in wavelength between an object and its background (the same center-surround trick we saw before), we can measure a property of the object—its color, which is independent of the illuminating wavelength. Thus, tigers are yellow all day long, giving a color-detecting animal like us a significant evolutionary advantage, as you can imagine, if you want to be able to detect them.

So what, then, is color? What we think of as color is actually a clever measure of the chemical properties of the surface of objects, as the absorption and reflectance of wavelength depends on the chemical composition of the object. This reflectance changes, as it should, when you change the wavelength of the illuminating light. By computing the color of an object, we cancel out the effect of the illuminating wavelength. With color perception we are, in a way, performing a spectroscopic analysis of the chemical properties of its surface, which does not vary and serves as an extremely sophisticated and useful tag for objects, since you cannot easily change your skin properties (except for chameleons and cephalopods). This has a lot of evolutionary advantages but also deep implications: If you think about it, color does not exist in the world. It is a computation that our brain makes, constructing it using wavelength information. That's why color is an example of Kantian perception: it is internally constructed. It's not made up; it absolutely applies to the world, as a tag of the chemical composition of the surface of objects, but it is something that our brain builds from the world.

The Blind Spot

If the idea that color is actually constructed by our brains seems weird to you, just look up some color illusions on the web. You will be amazed to realize how our perception of the color of an object can completely change if we muck around with its background illumination. But one could say that color illusions are abnormal situations not normally encountered in the world, so why worry? Well, I am afraid we do need to worry, because the fact that color computation is generated internally may not be an exception but the rule of most—perhaps all—brain computations. In the case of color, one can demonstrate clearly that our perception does not always correspond to reality but is instead generated internally. In case you need more convincing, let's look at another example of a Kantian perception, one that is constantly present in our vision yet one we never "see": our blind spot.

It turns out that our retinas have a spot (the blind spot) where the axons from retinal ganglion cells exit. And, because of geometric constraints, at the blind spot there is no space for photoreceptors so the light hitting it is not detected (hence the name "blind"). We actually have two blind spots, one in each retina, but because they are not in optical register, when the image of an object hits one, it also hits the normal part of the retina in the other eye, so we have no problem detecting objects that appear in front of one of the blind spots. But what would happen if you had only one eye? Let's try closing one eye and positioning an object, like your finger, so its image falls into the blind spot of the open eye. You actually don't see it! If you move the tip of your finger in and out of the blind spot, it disappears and reappears. I still remember the first time I tried this illusion and how shocked I was about how things appear and disappear in our vision. Try this yourself and have fun (figure 6.20).

Now comes the most interesting thing about what you "see" in the blind spot. The image that falls onto the blind spot is not black, dark, or empty. What you see in the blind spot resembles the neighboring areas of the visual field. In other words, your brain—and here I mean your visual cortex—is filling in the blind spot with a made-up image that resembles the area around it. It is a fake! The brain could have said, "Hey there,

6.20 The blind spot illusion. If you face this figure straight, cover your left eye and look at the cross, and then move your head closer or farther away from the page, you will notice that at some point, the eye disappears, replaced by a uniform background. This happens because the image of the eye falls into the **blind spot** of your right retina. If you move your finger into the position of the eye, you will notice that its tip disappears too.

you can't see anything there, so let's make it black or white." Instead, it chooses to make up an artificial perception that is similar to the surrounding territory, as if guessing what's out there. It's useful but, again, it's not real. It's made up, "constructed." Kant is probably smiling.

▼ RECAP

The neuroscience of vision is full of history, discoveries, concepts, and extraordinary phenomenology. How do we see? Operating over ten orders of magnitude in illumination intensity, the brain uses luminance and wavelength information from objects to calculate image contrast and builds images in a piecemeal fashion, creating sketches of the outline of objects. That happens with increasing sophistication along the visual pathway, from the retinal to the dorsal and ventral streams. Objects are recognized by their outlines, their movement is computed, and they are tagged with a color, and all of this is done at a distance without ever touching the object. And these computations are performed in an orderly fashion through maps of the world, of orientation, of direction of movements, and of categories of objects. We even have spectacular pinwheels of orientation selectivity mapped in our visual cortex.

An engineer could look at all this and say that it's straightforward and logical. Nature is using vision to measure the amplitude (luminance), position (shape), wavelength (color), and phase (motion) of objects, as visual primitives, and then combining those measurements in some abstract space to identify clusters that represent objects and their physical and even surface chemical properties.

Why do you want to do that? To predict the future. You need to know what's out there and how it moves so you can anticipate where it's going to be. To predict more accurately, you need to measure how things change so you focus on reflectance, differences, contrasts, and saliency and extract as much information about the object as you can. And if you don't happen to see the object because it falls in your blind spot, you just have to take a guess. It all makes good sense.

Further Reading

Dowling, J. E. 2012. *The Retina: An Approachable Part of the Brain*, rev. ed. Cambridge, MA: Belknap Press of Harvard University Press. A well-written introductory classic of the structure and function of the retina.

Hecht, S., S. Shlaer, and M. H. Pirenne. 1942. "Energy, Quanta and Vision." *Journal of General Physiology* 25: 819–40. In this landmark experiment, carried out in the building next to mine at Columbia, three biophysicists discovered that our visual system detects individual photons operating close to the physical limit.

Hubel, D. H. 1988. *Eye, Brain and Vision*. New York: Scientific American Library. A lucid account of the electrophysiological discoveries of a golden era of research in vision.

Sacks, O. 1999. *The Man Who Mistook His Wife for a Hat: And Other Clinical Tales*. New York: Harper Collins. A fascinating book by a uniquely perceptive neurologist, who regales us with stories from patients who have selective neurological deficits. You won't be able to put it down.

Sterling, P. 1990. "Retina." In *The Synaptic Organization of the Brain*, ed. G. M. Shepherd. Oxford: Oxford University Press. An excellent summary of the retina and its circuitry.

Wandell, B. A. 1995. *Foundations of Vision*. Sunderland, MA: Sinauer Associates. Another well-written book on the visual system, this time from a psychophysical angle.

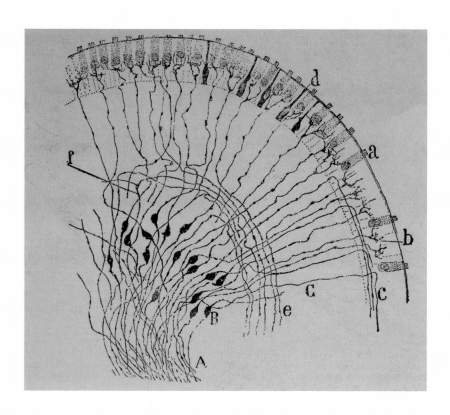

Section through the cochlea of a mouse. The cochlea, a marvel of biological engineering, has hair cells that transform sound into electrical signals. Courtesy of the Cajal Institute, "Cajal Legacy," Spanish National Research Council (CSIC), Madrid, Spain.

LECTURE 7: AUDITION

Close your eyes, prick your ears, and from the softest sound to the wildest noise, from the simplest tone to the highest harmony, from the most violent, passionate scream to the gentlest words of sweet reason, it is Nature who speaks, revealing her being, her power, her life, and her relatedness so that the blind person, to whom the infinitely visible world is denied, can grasp an infinite vitality in what can be heard.
— Johann Wolfgang von Goethe, *Theory of Colors* (1810)

OVERVIEW

In this lecture, we'll discover

- ▶ How the cochlea decomposes sound into frequencies
- ▶ How hair cells exquisitely detect movement
- ▶ How object location is computed by the brainstem
- ▶ How the cortex maps abstract sound properties

Sound

Sound conveys information from a distance about objects and their location, an obvious complement to vision and olfaction for a multipronged identification task. Let's first look at the physics. Sound is a series of compressions and expansions of waves that move through the air at 300 m/s.

It is created when an object vibrates in a solid, liquid, or gas, through which the sound travels by vibrating the molecules of the medium (figure 7.1). The movement of the molecules within a medium is determined by the spacing between them: sound travels faster through solid than liquid media, and faster through liquid than gas. Vibration consists of forward and backward movement of the object; this pattern creates a longitudinal wave, meaning that the media particles also move forward and backward in the same direction as the propagation of the wave, compressing them together and then spreading them apart, which is called *rarefaction*. We describe this movement as a wave, with high points representing compression of the particles and low points representing its rarefaction.

Sound, then, is just a wave, and like all waves, it can be characterized by its frequency, amplitude, and phase. So, as you can guess, the auditory system measures sound frequency, amplitude, and phase. How's that for good engineering? The frequency is what we refer to as the *pitch* of the sound and the amplitude, its *loudness*, and we use the phase to localize object sounds in space. The analogy to engineering goes further: because

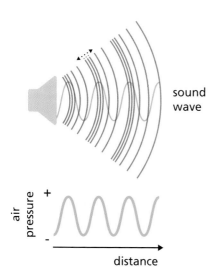

7.1 Sound is a vibration in an object, transmitted as longitudinal pressure waves.

sounds waves are often complex, our auditory system decomposes them into mixtures of sinusoidal waves. Why? According to the Fourier theorem, you can mathematically decompose any complex wave into a sum of sinusoidal waves. So it looks like evolution discovered this theorem before we did!

If you are not yet impressed with what's going on in our ears, let me give you two more facts. The first is how the auditory system solves an enormous problem: the impedance mismatch. It turns out that when a pressure wave changes media with different impedances (i.e., different propagation speeds)—for example, as it goes from the air into bone—it is reflected back. Thus, airborne sound should bounce off our bodies and should never penetrate our ears. But our body compensates for this huge loss in sound propagation as it enters our ears by amplifying the sound wave with a clever design and using mechanical levers.

The second "wow" fact is that our auditory system is sensitive over sixteen orders of magnitude—it beats vision! We can hear both incredibly loud and incredibly soft sounds. How soft? Our hair cells can detect movements as small as the diameter of an individual atom! This is right up to the physical limit of sound. It's not the first (or last) time that we will discover that evolution has reached the physical limits. As we saw in the last lecture, photoreceptors can detect light intensities as small as individual photons, and, as we will see in the next lecture, olfactory systems in some animals can also detect the presence in air of individual molecules. This is a serious optimization job.

Evolution has pushed sensory systems to the physical limit of detection. This is as good as it gets. This teaches us an important lesson: if the sensory systems are optimized to the physical limit, it makes sense to expect that the neural circuits that analyze this information in the brain also should be close to the limit of what is computationally possible. Otherwise, why bother having such a perfect front end to the brain if you are going to throw away that precision later on, when you use sensory information to compute or make a decision? Imagine the secrets that are waiting for us up in the brain.

The Ear

How do we hear? How does the auditory system perform these amazing feats of engineering? It's all in the ear. The mammalian ear is divided into three parts: the outer, middle, and inner ear (figure 7.2). The *outer ear* includes the pinna (what we normally call the ear, and what we can see from the outside) and the ear canal. Have you ever wondered why our ears are so odd-looking? Well, it seems that the particular shape of the pinna is critical for sound localization, because it captures sound differently at different angles depending on where the sound is coming from. If we feed sound directly into the ear canal, we lose a lot of sound localization. Also, the shape of the pinna, which resembles a bullhorn, is critical because as it funnels sound to the ear canal, it amplifies it approximately a hundredfold.

After the sound is collected in the pinna, it travels through the ear canal and hits the tympanic membrane (our eardrum) in the *middle ear*,

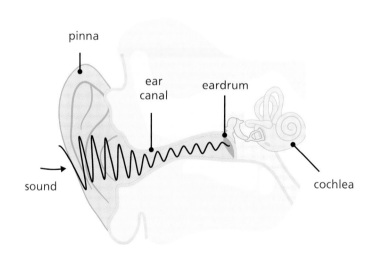

7.2 The **outer ear** gathers sound waves and amplifies them, delivering them to the eardrum in the middle ear, which passes them to the cochlea.

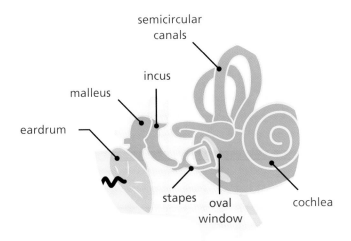

semicircular
canals

incus

malleus

eardrum

stapes

oval
window

cochlea

7.3 The three bones of the **middle ear** act as a lever, amplifying the sound amplitude and delivering it to the inner ear.

with its three tiny bones, the smallest in the body: the malleus, incus, and stapes (figure 7.3). These *ossicles* act as levers, transmitting mechanical energy from the vibrating tympanic membrane and thus provide a further (up to two-hundred-fold) amplification to the sound waves, which have been greatly diminished as they enter the body. Since the inner ear is fluid-filled, the sound, which has been moving through a low-impedance medium—air—in the inner canal, now enters a higher-impedance liquid medium, one that allows faster speed. The combined one-hundred- and two-hundred-fold amplification that is carried out by the outer and middle ear, respectively, solves the impedance mismatch problem. Otherwise practically no sound would make it in.

After leaving the middle ear, sound pressure waves enter the inner ear, which is formed by the cochlea and the semicircular canals (figure 7.4). The tiny bones in the middle ear press on a membrane in the cochlea called the *oval window*. The cochlea itself is a bony snail-shaped tube; it is filled with fluid and has at its center a thin elastic membrane with

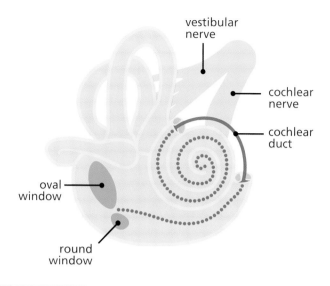

vestibular
nerve

cochlear
nerve

cochlear
duct

oval
window

round
window

7.4 The **inner ear** contains the cochlea, a rolled-up tube that measures sound frequency.

hair cells that runs along its entire length. It is here that the mechanical energy of the sound pressure waves is transformed into electrical signals.

The Cochlea

Now, why is the cochlea a tube? If it were unrolled from its snail shape, it would look like a tongue about 30 mm long, getting wider at its tip (figure 7.5). This shape has a major influence on hearing. It is tonotopic in organization: high-frequency waves vibrate the narrower base of the cochlea, while lower-frequency waves vibrate the wider tip. This makes sense if you think about it. The smaller an object is, the faster it can be moved, and vice versa, similar to how musical instruments work. For example, the lower sounds of a piano are made by longer strings and higher sounds by shorter strings. In fact, Hermann von Helmholtz, a German polymath who discovered this phenomenon, compared the cochlea to a piano, a fact confirmed by Georg von Békésy,

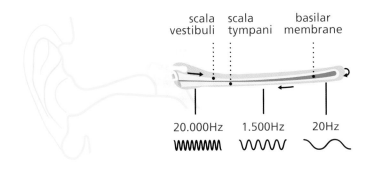

7.5 The **cochlea** vibrates at different sound frequencies in a tonotopic fashion.

another extraordinary scientist. And, just like a piano, the scale of the cochlea is logarithmic, which allows animals to detect a broad range of frequencies. For example, humans hear typically between 20 Hz and 20,000 Hz, and speech usually falls within 100 Hz and 200 Hz. In fact, the range of detectable frequencies is partly determined by the physical dimensions of the cochlea: larger animals can hear lower frequencies and smaller animals, higher ones. Isn't this match between form and function beautiful?

How does the cochlea work? Let's examine a cross-section (figure 7.6). This tube is composed of three parallel chambers that run all the way through it: the scala tympani (connected to the tympanic membrane) in the bottom, the scala media (in between), and the scala vestibuli on top, connected to the vestibular duct. The middle compartment, the scala media, is really interesting, as it is filled with the endolymph, which resembles the intracellular medium in ionic composition (with a lot of potassium and little sodium, if you remember from the third lecture) and bathes the top of special neurons, called *hair cells*, located on the floor of the scala media on a structure called the *basilar membrane*. Meanwhile, the scala tympani and the scala vestibuli contain perilymph, which is rich in sodium but contains little potassium, as is typical of all extracellular fluid in the body. The difference in ionic composition between these chambers is key, as we will see.

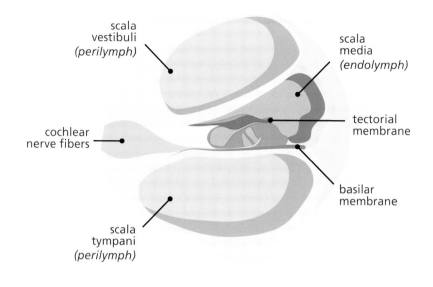

scala
vestibuli
(perilymph)

scala
media
(endolymph)

tectorial
membrane

cochlear
nerve fibers

basilar
membrane

scala
tympani
(perilymph)

7.6 The **cochlea** has three fluid-filled tubes (scalae), separated by two semirigid membranes.

Hair Cells

The structure of the cochlea appears quite complicated. How exactly does the cochlea detect sound? That's where the hair cells come in. They got their name because they have stereocilia at their tips. Stereocilia are a bundle of hairlike projections made from cross-linked actin and are arranged in ascending order of height, like treetops up a mountain slope. The tops of stereocilia touch the tectorial membrane, a bony structure on the scala media that is attached on one end to the cochlea, with its other end free floating (figure 7.7). The tectorial membrane vibrates due to an incoming sound wave, which then wiggles the stereocilia back and forth. That wiggling results in hair cells' depolarization and subsequent transmission of electrical signals to the brain.

Let's talk about the details of this process. It turns out that each stereocilium is linked by a fine thread—the so-called tip link—to its neighboring

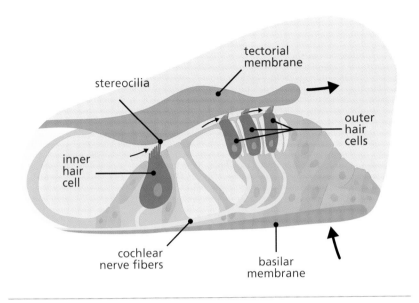

7.7 The top of the **hair cells** (blue) are anchored in the tectorial membrane and are displaced back and forth due to the sound-induced vibrations. Hair cells transduce that movement into electrical signals sent to the brain.

cilium. So one end of a tip link is attached to the side of the stereocilium and the other end to the top of a channel of its neighbor (figure 7.8). Yes, I meant to say this, to the top of a channel. A single channel. Talk about nanotechnology! The idea is that when the stereocilium moves in one direction (because it's carried over by the tectorial membrane, reflecting the sound wave), it pulls the tip link taut and opens the channel, depolarizing the cell. But if it moves in the opposite direction, the tip link slackens and the channel closes (figure 7.9). That's how you can transform a movement into an electrical signal—by opening and closing these channels, due to the tightening and slackening of tip links. The whole point is for the sound to vibrate the cochlea in order to move the stereocilia back and forth, causing the hair cell to become depolarized or hyperpolarized depending on the motion.

I have to confess that since I first heard about the tip link hypothesis, I continue to have trouble believing it. How can you build such tip links

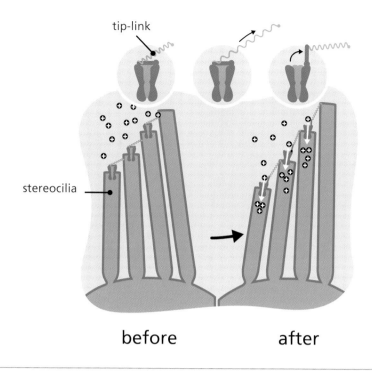

tip-link

stereocilia

before   after

7.8 Tip links are filaments of hair cells sterocilia that link the top of a channel with the neighboring cilium, opening the channel when the bundle of cilia is displaced.

and attach them so precisely? How do you keep them taut? How is the length and position of each cilium controlled to ensure that the tip links work? How can a channel detect the movement of the diameter of single atoms? How can a hair cell detect the opening of a single channel? How do you arrange all the stereocilia in a neat, orderly fashion? How do you make all the tip links and channels in different cilia work in concert, as opposed to at odds with each other? How, how, how . . . ? If Darwin had known about tip links, he would have replaced the retina with the cochlea in that famous passage we saw in lecture 6.

We are still not done with surprises. Let's look at how hair cells generate an electrical signal. As you may remember, the stereocilia are located in the scala media, bathed in the K^+-rich endolymph, so when the tip

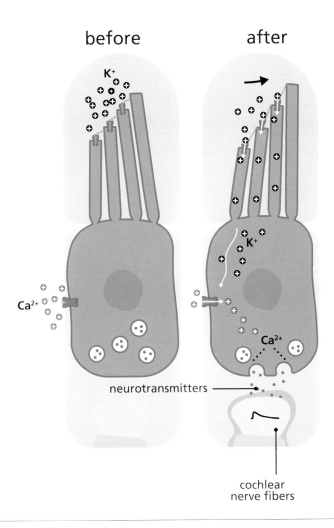

before **after**

7.9 Hair cells are depolarized when K⁺ flows through the channels opened by tip links. The depolarization activates calcium channels that release neurotransmitter, depolarizing the nerve terminal.

link channel opens, K⁺ (rather than the usual Na⁺) comes into the cilia and depolarizes the hair cell (figure 7.9). This depolarization spreads through the membrane of the hair cell and opens voltage-sensitive calcium channels at the bottom of the hair cell, which triggers transmitter release, depolarizing the terminal of the afferent nerve of the cochlear nerve which then carries all auditory information to the brain. Overall, the mechanism of mechanoelectrical transduction is actually really good. How good? Hair cells in the cochlea can respond to a 3Å (three Angstroms) movements of their cilia. That's about the diameter of a midsize atom. Here again we have evolution at its finest: our neural hardware is good to the physical limit!

One last amazing fact about the cochlea: we talked about Helmholtz's piano, and how different parts of the cochlea resonate to different frequencies of sound, reflecting the different widths of the basilar membrane. Well, it seems that the hair cells located in different parts of the cochlea also differ in electrical properties, so they match the frequency of the spot along the cochlea where they are located. This is apparently achieved by differences in the density and kinetics of potassium channels; faster channels are expressed in hair cells transducing faster frequencies of sound, and slower channels are found in lower-frequency cells. To achieve this, you need to precisely regulate the expression of potassium channels along the length of the cochlea. Impressive! Darwin would totally give up!

Before we leave the cochlea, I should mention that there are two types of hair cells: the inner and outer hair cells (notice them in figure 7.7). The ones we have been talking about are the *inner* hair cells, and they are actually a minority. Most hair cells are actually *outer* cells, which, instead of transducing mechanical stimuli and sending them via afferent fibers to the brain, receive efferent axons from the brain! In other words, they work backward, transducing electrical signals into mechanical movements. Thus, outer hair cells are like the brain's minions in the cochlea and are thought to serve as mechanical levers that can push the membrane up or down and thereby help maintain the tension of the tectorial membrane. In fact, it turns out that, if a microphone is placed in the ear, it detects spontaneous sounds coming from the ear at particular frequencies. In other words, it seems that the brain can run the cochlea backward and generate sound with the outer hair cells, while it uses the inner hair cells to detect it. The purpose of this is all very mysterious. Somehow the brain is tinkering actively with the cochlea, perhaps providing feedback control and support needed to keep the membrane at the sensitivity needed to pick up a particular frequency—or, as we have discussed regarding the thalamus, actively modulating, amplifying, or blocking particular auditory signals that the brain is interested in listening to. Again, here we see the same theme as in vision: sensory information, and thus perception, being controlled by the brain.

Brainstem

After sounds have been transformed into electrical signals by the cochlea, axons carry that information as patterns of action potentials in the cochlear nerve to the brain. This ascending auditory pathway first stops in the cochlear nucleus and the olivary body, which sit in the medulla of the brainstem, before projecting to the thalamus and the cortex (figure 7.10).

We only understand a little bit what these brainstem nuclei do with this sound information, but what little we do understand is very impressive, confirming the suspicion that the precision in design of the inner workings of brain circuits matches that of the sensory organs.

One thing we know is that the cochlear nucleus in the medulla has tonotopic (i.e., ordered by frequency) maps of sounds and the olivary body helps to localize sound in space. Sound localization is done with a remarkably clever circuit trick that takes advantage of the time and intensity differences in sound arrival to our ears (figure 7.11). Because our ears are separated in space by our head, any sound from an object will hit one ear before the other, unless the sound source is exactly in front of us. Thus, for most sounds, there is a small time difference in when sound arrives at one ear versus the other. This *interaural* (i.e., between-ears) time difference is tiny. Let's say that our ears are 20 cm apart, and assuming a speed of sound in air of 300 m/s, sound from an object located on our left will hit the left ear approximately 600 micros before the right ear. Well, it turns out that we can detect interaural time differences as small 10 micros—that is, sixty times faster than we should be able to do based on the physics of the sound and dimensions of our heads! How can you beat physics? With neural circuitry. This is where the olivary nucleus comes

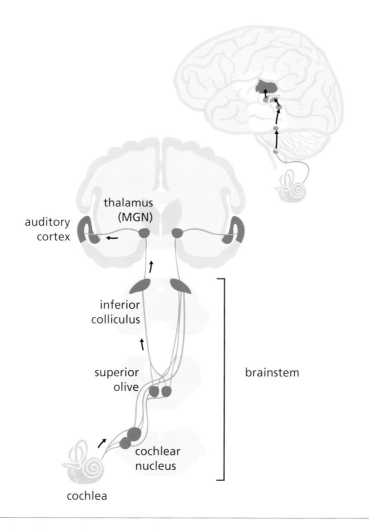

7.10 The **auditory pathway** ascends from the cochlea to a series of nuclei in the brainstem, reaching the thalamus and auditory cortex.

in. Its neural circuitry is arranged like this: there are two parallel sets of axons, one coming from the left ear and another one coming from the right (figure 7.11). These axons contact the same set of postsynaptic neurons, which are neatly arranged in a line from left to right. Now, because of the delay in axonal transmission, when the sound comes from the left ear, the

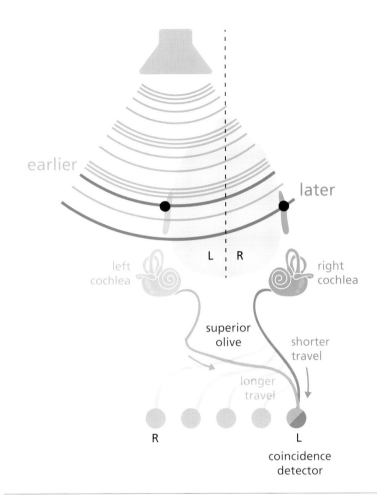

7.11 Sound localization is computed by the olivary nucleus by taking advantage of different delays of sound arrival to the two ears, and mapping those delays to a row of cells, where each of them is activated by a particular interaural difference that corresponds to a horizontal position in space.

neurons that are closest to the left ear will be activated ahead of the neurons that are farthest away from the left ear. And the same thing happens with neurons closest to the right ear when the sound comes from the right side.

Now consider a situation in which the sound source is exactly in the center, directly ahead of us. The neurons located in the middle of the row will be hit simultaneously by activity coming from the left and right ear, and bingo! That neuron will fire as a detector that codes the coincidence of sounds stemming from objects that are directly in front of us. Next, think about what happens if we move the sound source to one side. A particular neuron in the olivary nucleus will respond, because it receives axons from the left and right ear that have exactly the right sets of complementary delays. With this design, each neuron in this row in the olivary nucleus responds to sound sources located at a particular position in the horizon. We have just built ourselves a detector of sound sources, taking advantage of the interaural time differences and cleverly matching them to the axonal delays. Another way to put it is that we have computationally extracted a space map using interaural delays. We have built an internal map of the outside world.

In addition to interaural *time* differences, another part of the olivary nucleus uses interaural *intensity* differences to localize the sound. Again, we take advantage of the physics. Think of the head as casting a sound "shadow," so that if an object is closer to one ear, it will be louder there than in the other ear. By using inhibitory axons from one side of the olivary nucleus to the other, each ear tries to shut off the incoming stimulus from the other ear. Incidentally, this winner-take-all neural circuit resembles the *half-center*, cross-inhibition model of Graham Brown we mentioned when discussing the spinal cord. With this contralateral inhibitory circuit, our brainstem amplifies interaural intensity differences and uses that information as a second strategy to help localize the sound in the horizontal plane. Now, what about sound localization in the vertical plane? We don't quite know how this is done, but the asymmetric shape of the outer ears, and the peculiar folding of our pinnae, appears to help, probably by creating a differential amplification of sounds that come from the top or the bottom.

From the olivary nucleus in the medulla, most of the auditory pathway axons connect to the inferior colliculus, which sits under the superior colliculus in the midbrain (also within the brainstem). Using the sound localization information from the olivary nucleus, the inferior colliculus is involved in building more refined spatial maps of sound information, whereas the superior colliculus is involved in building visual maps of the world. These two sets of auditory and visual maps are actually in register (or aligned), so an object located in a particular point of the visual field stimulates a neuron of the visual map in the superior colliculus that is connected to the neuron in the auditory map in the inferior colliculus that responds to sound emanating from that same position. What do you do with all these maps? You use them for vision and to orient reflexes, like turning your head to the source of a loud noise or a light.

Auditory Thalamus and Cortex

Let's move farther up the brain (figure 7.10). The auditory pathway continues to the medial geniculate nucleus of the thalamus (i.e., the auditory thalamus) and from there to the primary auditory cortex, replicating what we learned for vision. And just as with vision, we are not sure exactly what happens in the thalamus. In fact, the more we climb up the hierarchy, the less we understand. In bats, which, as mammals, are actually close relatives to us, neurons in the auditory thalamus respond to frequency-modulated (FM) sounds. FM sounds are important for the bat's echolocation, and we actually make such sounds when we speak. FM sounds are the consonants in our language, as opposed to vowels, which have stable frequencies.

In addition to building receptive fields of increasing complexity, it is likely that the auditory thalamus serves as an attentional filter, so only certain auditory signals that are deemed important by the cortex are allowed entry. You would expect this effect to be quite important, since the auditory pathway has a very strong efferent (top-down) pathway, reaching all the way down to the cochlea, whereas the visual pathway top-down control ends in the thalamus and does not reach the retina.

The primary auditory cortex is in the temporal lobe (figure 7.12) and is again arranged in a tonotopic fashion, with different frequencies mapped systematically. This arrangement is very similar to the columnar organization of the visual cortex, except that here we are dealing with sound frequencies as opposed to orientations. And, just like the visual cortex,

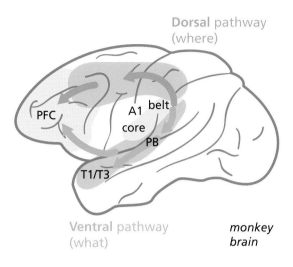

Dorsal pathway
(where)

PFC

A1 belt
core
PB

T1/T3

Ventral pathway
(what)

monkey
brain

7.12 Sound information reaches the **primary auditory cortex** (A1, yellow), where it is processed and then routed to the dorsal (red) and ventral (green) sensory streams, which converge in the prefrontal cortex (PFC, blue).

the auditory cortex contains many types of maps, with ocular dominance columns for vision replaced by maps for audition with binaural or monaural cells, which keep track of which ear the sound is coming from. There are also maps for measuring loudness and probably other sound features yet to be discovered.

In a way, sensory cortices all work the same, with a combination of lower-level features, whether auditory or visual, building more sophisticated receptive fields and analyzing more complex stimuli, similar to the simple and complex cells from V1. Given that sound is a temporally modulated signal, a lot of processing in the auditory cortex probably involves detailed temporal sequences of stimuli, highlighting the idea that the cortex is a temporal machine.

The primary auditory cortex is surrounded by an *auditory belt* of secondary areas, connecting to the rest of the cortex in a manner similar to the visual cortex, with a where pathway for orientation and a what pathway to decipher the identification of the sound. And just as secondary and tertiary visual areas show more and more specialized properties, the secondary and tertiary auditory cortices feature sections that are cued for different types of sound, abstracting features from the sensory world. Cortex is cortex, no matter what function it serves, and its different parts likely apply the same algorithms and computational principles to different types of sensory stimuli. Given the importance of speech for humans, we probably have cells that are sensitive to vowels, consonants, and different combinations of them forming words or sentences. In fact, the detection of speech is a phenomenal computational problem, one that has become a cornerstone of computer science. But that's another story.

▼ RECAP

The auditory system is yet another engineering marvel with a beautiful, unparalleled match between form and function. It takes advantage of the physics of the head and the ear and a system of mechanical levers to perform major amplification of sound, solving the impedance mismatch problem.

In the cochlea, the auditory system performs an ordered mechanical Fourier-like decomposition of sound into its fundamental frequencies. Then come the hair cells, which transduce sound into electrical signals with atomic sensitivity. This front end of the auditory system, optimized by evolution to the physical limit and with a dynamic range of twelve orders of magnitude (120 decibels), sends, in parallel channels, amplitude (loudness), phase (timing), and frequency (pitch) information to the brainstem, thalamus, and cortex. There, elegant and powerful algorithms appear to be at work to analyze the position and nature of the sound source, build auditory maps of our environment and tonotopic maps of its sounds, extract sound signals from background noise, and decode biologically meaningful sounds like echolocation or speech. Once information reaches the cortex, we become conscious of it, so we can combine it with all other sensory information about the stimulus.

Finally, every step of the auditory pathway is likely under active top-down control by the cortex to tune the system and optimize its performance, selecting the biologically relevant information from the cacophony of noise in the world and likely helping to construct the auditory stimulus and an auditory model of the world in a Kantian manner.

Further Reading

Békésy, G. 1960. *Experiments in Hearing*. New York: McGraw-Hill. This book, by an early pioneer, describes the amazing discovery of the mechanical decomposition of sound by the cochlea.

Cahan, D. 2019. *Helmholtz: A Life in Science*. Chicago: University of Chicago Press. A biography of the nineteenth-century polymath who by himself opened up the study of the auditory system as well as many other fields.

Crawford, A. C., and R. Fettiplace. 1981. "An Electrical Tuning Mechanism in Turtle Cochlear Hair Cells." *Journal of Physiology* 312: 377–412. This paper reports the remarkable discovery that hair cells are electrically tuned to oscillate at frequencies that correspond to their cochlear positions.

Hudspeth, A. J. 2008. "Making an Effort to Listen: Mechanical Amplification in the Ear." *Neuron* 59, no. 4: 530–45. A review of mechanotransduction by the hair cells, with beautiful biophysics and exquisite experiments by another pioneer in the field.

Suga, N. 1990. "Bisonar and Neural Computation in Bats." *Scientific American* 262: 60–68. A fascinating look into the function of the auditory thalamus of bats.

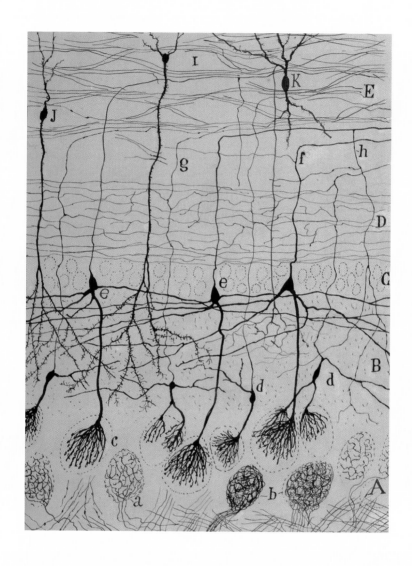

Section from the olfactory bulb of a cat. Olfactory information is elegantly sorted out at the bulb before it gets sent to the cortex. Courtesy of the Cajal Institute, "Cajal Legacy," Spanish National Research Council (CSIC), Madrid, Spain.

LECTURE 8: OLFACTION

No sooner had the warm liquid mixed with the crumbs touched my palate than a shudder ran through me and I stopped, intent upon the extraordinary thing that was happening to me. An exquisite pleasure had invaded my senses, something isolated, detached, with no suggestion of its origin. And at once the vicissitudes of life had become indifferent to me, its disasters innocuous, its brevity illusory—this new sensation having had on me the effect which love has of filling me with a precious essence; or rather this essence was not in me it was me. . . . Whence did it come? What did it mean? How could I seize and apprehend it? . . . And suddenly the memory revealed itself. The taste was that of the little piece of madeleine which on Sunday mornings at Combray (because on those mornings I did not go out before mass), when I went to say good morning to her in her bedroom, my aunt Léonie used to give me, dipping it first in her own cup of tea or tisane. The sight of the little madeleine had recalled nothing to my mind before I tasted it. And all from my cup of tea.

—Marcel Proust, *In Search of Lost Time* (1913)

OVERVIEW

In this lecture, we'll discover

- ▸ How olfactory receptors transduce odorant molecules into electrical signals
- ▸ How the brain uses a combinatorial olfactory code to detect objects
- ▸ How odors are constructed by the brain
- ▸ How taste follows a labeled line code

Odors

Smell is the detection of odors. But what actually is an odor? Odor molecules, or *odorants*, are volatile organic molecules, often lipophilic (not water soluble), that are emitted mostly by organic matter—that is, living organisms (figure 8.1). In other words, smell is a sensory system built to detect the presence of animals or plants. This ability is critical for survival, food detection, mating, and social relations. Smell is also "old news" in terms of evolution. The detection of important chemicals in the air and the water is an ancient sensory system, going all the way back to bacteria.

For terrestrial animals like ourselves, odorants are volatile and are transmitted by air currents in odor plumes. Air mixing and its dynamics are complex, so the olfactory system does not have the luxury of detecting objects at the speed of light, as the visual system does for photons. Instead, it has to measure odorants slowly and carefully in every sniff we take. Nevertheless, as in the visual and auditory systems,

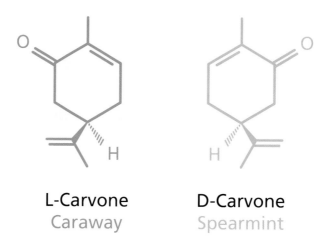

L-Carvone
Caraway

D-Carvone
Spearmint

8.1 Most **odorants** are lipophilic organic molecules with smells that correspond to their particular chemical structure.

nature has engineered the olfactory system to the physical limit: animals like moths can detect individual molecules of pheromones and change their behavior accordingly! Think about it: not only can a single molecule floating in the air be detected, but the brain knows about it and generates an appropriate behavior! Amazing detection abilities! Again, the brain uses similar principles for olfaction as it does for other sensory systems: parallel pathways for specialized measurement of stimulus properties and top-down control to "construct" perceptions. If you dig deeper, you'll find all the principles of neuroscience here: ensembles, hierarchy, wiring, maps, control, learning, and optimization. The olfactory system is a wonderful self-contained microcosm of the brain.

We often talk about smell and taste together, because they are very similar senses: both are specialized for detecting chemicals. In fact, they are in physical contact through the pharynx; and most of the taste of the food actually comes from the way it smells. In other words, you smell the food as you are chewing it. So if you catch a cold (or COVID-19, as just happened to me) and the virus wipes out your olfactory mucosa, you lose not only your sense of smell but most of your taste, too.

Olfactory System

Let's start with olfaction. How do we smell? Olfaction is a two-step job: it enables us to detect odorants and decode smells, their mixtures. Detecting is done by the nose and decoding by the brain. Let's dive in: In mammals, the olfactory pathway (figure 8.2) starts in the lining of the nose, the olfactory mucosa, which has an epithelium with sensory neurons that project to the olfactory bulb. The bulb then has neurons that project to the cortex and many parts of the brain.

Smell is peculiar; it's the only sense that doesn't go through the thalamus before entering the cortex. No one knows why, although some suggest this is due to the ancient evolutionary origin of olfaction. Maybe, maybe not.

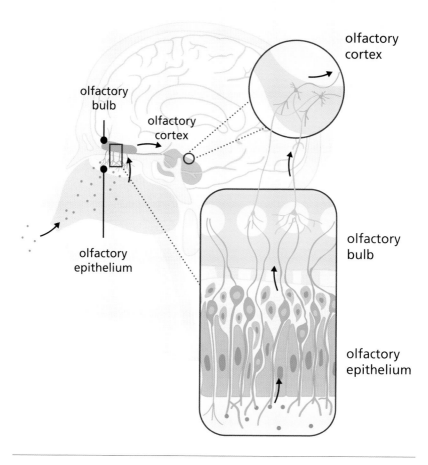

8.2 The **olfactory system** detects odors in the olfactory epithelium and sends this information to the brain for smell identification.

Olfactory Receptor Neurons

Our journey starts in the olfactory epithelium, at the top of the nasal cavity, where smell detection occurs (figure 8.3). It's probably not news to you that the nose is full of mucus. Did you ever wonder why? Mucus, secreted by specialized cells in the olfactory epithelium, traps and concentrates

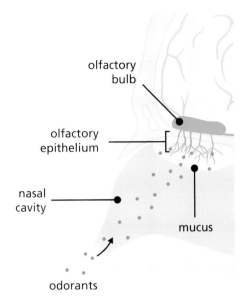

olfactory
bulb

olfactory
epithelium

nasal
cavity

mucus

odorants

8.3 Odorants are breathed into the nose and are trapped by the mucus in the **olfactory epithelium**.

odor molecules, both through its particular chemical properties and via specialized binding proteins. This is a key job, given that odorants are volatile and often lipophilic. I hope from now on you will treat your mucus with some respect!

Once odorants are trapped, the mucus brings them into contact with the cilia of the olfactory receptor neurons (ORNs), which sit on the surface of the olfactory epithelium (figure 8.4). These cilia, which project into the nasal cavity, have olfactory receptors, fascinating proteins that specifically bind odorants. Olfactory receptors (ORs) are illustrious members of a family of G-protein-coupled receptors (or GPCRs) with seven transmembrane regions (figure 8.5). We have about three hundred fifty ORs, which share the same structural scaffold but differ in the amino acid sequence that builds their binding pockets for odorants, enabling recognition of diverse chemicals. This can help explain how we detect thousands of odors.

Just as different antibodies detect different antigens through their different binding pockets, different odorants bind to specific ORs. And, just

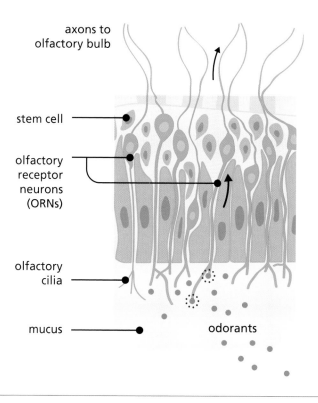

axons to
olfactory bulb

stem cell

olfactory
receptor
neurons
(ORNs)

olfactory
cilia

mucus

odorants

8.4 The olfactory epithelium has **receptor neurons** that detect odorants and generate electrical signals sent to the olfactory bulb for processing.

like antigens can bind to different antibodies, each odorant can bind to several ORs, with higher and lower affinity, as it fits binding pockets more or less snugly. Thus, as a group, ORs form a series of molecular detectors that, together, can bind to essentially any molecule. In addition, just like the B-lymphocytes that make antibodies, olfactory neurons not only choose to express only one OR out of all those available in its genome but also display allelic exclusion, a mechanism by which only one allele of a gene is expressed while the other allele is silenced. (Remember that each cell has two alleles for each autosomal gene.)

The bottom line is that from the three hundred fifty possible ORs present in our genome, each neuron expresses only one, so the message it sends to the brain is specific. This strategy of one ORN = one OR also ensures that all ORNs are different from one another, covering all the chemical space of potential odorants. Nature is using the same trick as in the immune system, which is also tasked with detecting any possible antigen: to make different types of binding pockets in different receptors and express them selectively in individual cells.

Olfactory Transduction

What happens next is the transduction of the chemical signal—the odorant—into an electrical signal that gets sent to the brain. Transduction occurs on the membranes of the cilia of the ORNs. Once an odorant binds to the receptor,

8.5 Olfactory receptors are G-protein-coupled receptors with seven trans-membrane helixes.

it triggers a second-messenger cascade, as any good GPCR is supposed to do (figure 8.6). This process begins with the activation of a G protein (G-olf), which serves to signal that the odorant binds to the receptor. G-olf then activates an adenyl cyclase (ACIII), an enzyme which makes a lot of cyclic adenosine monophosphate (cAMP). cAMP then floods the neuron and activates a cation channel that brings sodium and calcium into the cell. As good cations, they are positively charged and depolarize the neuron. In addition, calcium also triggers a series of secondary events to shape that depolarization.

Just like phototransduction, olfactory transduction likely uses this multistep biochemistry to amplify the signal. And this all happens on the plasma membrane for speed, as every diffusional step is faster in two dimensions. If the olfactory neuron becomes sufficiently depolarized, it fires action potentials and sends that information up into the brain.

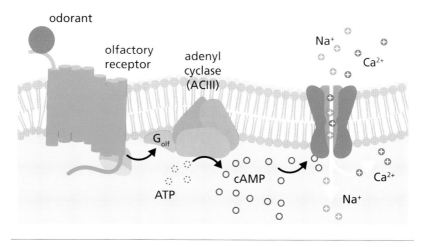

8.6 Binding of odorants to olfactory receptors triggers a biochemical cascade that generates **cAMP**, opens channels and depolarizes the ORN.

Olfactory Bulb

The first stop in the journey to the brain is the olfactory bulb, where axons from ORNs project to round structures called *glomeruli* (figure 8.7; see also Cajal's beautiful drawing at the beginning of the lecture). Here they make contact with dendrites from mitral cells, neurons that project into the cortex.

Something remarkable happens at the bulb. As you now know, each ORN expresses only one of the three hundred fifty ORs. But there are a lot of ORNs, many more than ORs, so many ORNs happen to express the same OR. Now, these olfactory neurons that express the same OR are all scattered in the nose, but, amazingly, their axons all converge onto the same glomerulus, so each glomerulus receives information from only one type of neuron expressing one type of receptor. Therefore, a given odor, which, let's say, would perhaps bind to a few ORs, will activate a subset of ORNs, which will then activate a selective subset of glomeruli. Thus, through the filter of ORs and ORNs, we have transformed a chemical signal into a spatial map of active glomeruli. We have encoded odorants into a map.

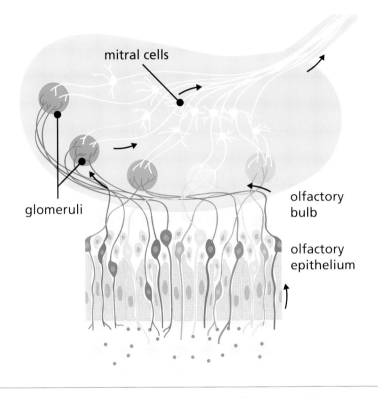

8.7 The **olfactory bulb** has glomeruli that receive axons from ORNs, which contact dendrites from mitral cells that project to the brain.

Now, if you want to decode what you're smelling, all you need to do is read out that map, the pattern of activation of all glomeruli, since each particular pattern corresponds to one odor. To read out those patterns is presumably the role of the cortex. This is a beautiful example of a *combinatorial* code, in which a combination of elements (in this case, active glomeruli) encode a signal. Nature is essentially building a matrix where each row is an odorant and each column, an OR (figure 8.8). This matrix serves as a sort of dictionary in which you can look up the combination of activated columns and decipher the odorant that triggers it. This decoding is done by the cortex, but let me point out that it does not have the benefit of having a built-in dictionary in a shelf to

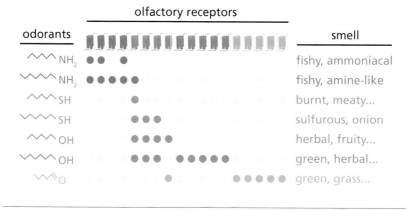

odorants	olfactory receptors	smell
NH₂	● ● ●	fishy, ammoniacal
NH₂	● ● ● ● ●	fishy, amine-like
SH	●	burnt, meaty...
SH	● ● ●	sulfurous, onion
OH	● ● ● ●	herbal, fruity...
OH	● ● ● ● ● ● ● ●	green, herbal...
O	● ● ● ● ● ●	green, grass...

8.8 Odorants are encoded by the **combinatorial** activation of ORs. By reading the combination of activated ORs, one can decode which odorant was present.

consult; it has to assemble this dictionary by itself, and it does so probably continuously.

But before we move up to the cortex, in case you are still not impressed, let me point out a couple more remarkable things about the olfactory bulb. It turns out that each bulb has not one but two glomeruli that receive axons from the neurons that express the same OR. Why? No idea. And to make it even more mysterious, in our two olfactory bulbs, one in each side of the brain, and the position of these two glomeruli is symmetrical in each of the bulbs! I feel sometimes that nature is thumbing its nose at us.

Olfactory Cortex

So we now have a spatial map of glomeruli activation to decode. How does the brain do this? Axons from the mitral cells project to the olfactory cortex, also known as the *pyriform cortex*. These projections are widespread: a given mitral axon contacts many cortical neurons, and a given cortical neuron receives axons from many different mitral cells (figure 8.9). Now, if you think about it, this means that each cortical neuron will be

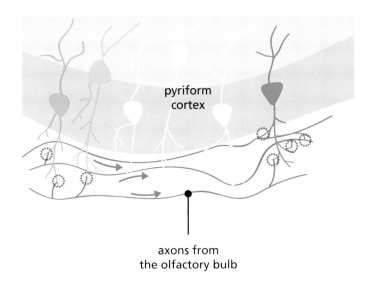

pyriform
cortex

axons from
the olfactory bulb

8.9 In the **olfactory cortex**, each mitral cell's axons contact many cortical neurons, distributing widely the signal detected by each OR.

contacted by a random subset of axons, each relaying the information from a subset of ORs. Why would you want to do this, and scramble all the information from each OR that you had neatly separated into different glomeruli? Well, precisely for this purpose, to scramble the information. In fact, the more you scramble it, the better.

This all has to do with both feedforward and feedback neural networks again, as we will see in a moment. Here's the logic: cortical neurons receive inputs from a combination of different glomeruli, and a given cortical neuron will only fire if those inputs happen to fire at the same time. Thus, that neuron becomes active if, and only if, a particular combination of glomeruli has turned on. In fact, didn't we say that a given odor activates a combination of glomeruli? Bingo! That cortical neuron is detecting a particular odor! It's reading the spatial code of the bulb and picking up one particular combination. Now that's when the other cortical neurons come in: each of them responds to one particular odor, and all of them together will decode all odors.

This decoding strategy is nothing new: we saw it in lecture 5 in the feed-forward neural networks. The olfactory cortex works as a perceptron that identifies odors; and the cortical neuron fires if a given combination of inputs is active, acting as an AND gate. That's how you build input selectivity. And if we add synaptic plasticity to the mix, we can play with the strength of the connections and build a "dictionary," an odor-decoding circuit that can learn and adapt to new odors! The olfactory system is revealing before our eyes the basic principles of neural networks. Isn't this match between the structure of the anatomy and the computational logic just plain beautiful?

Mitral cell axons project to places in the brain other than the olfactory cortex (figure 8.10), including the entorhinal cortex and the amygdala,

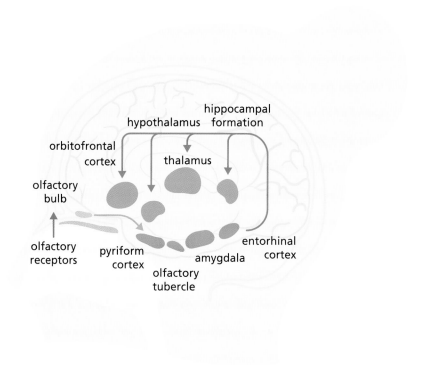

8.10 Olfactory processing occurs in many areas, including the amygdala, hippocampus, hypothalamus, and frontal cortex.

the "Grand Central Station" for fear. It turns out that olfactory responses can be innate or learned. For example, mice instinctively know to run from the smell of foxes, and we also learn to like or dislike smells during our lives. We suspect that the amygdala mediates innate olfactory responses and the cortex mediates learned responses. From these first stations in the olfactory pathway, second-order neurons then project widely throughout the brain to the frontal cortex (center of higher cognition and motor planning), thalamus (for attention), and hypothalamus (for emotions and hormones).

There are also significant projections to the hippocampus, which is involved in the storing of long-term memories, as we will see in lecture 17. The current idea is that these olfactory memories are attractors, stable or semistable states of activity in a recurrent neural circuit. In fact, you can even activate olfactory memories internally, without an odor, either by recalling or sometimes as olfactory hallucinations in epileptic seizures. Feedback networks, as you know, not only can encode memories as circuit states but are also endowed with pattern completion properties. This nicely explains Proust's madeleine recall. Who knew that Proust was a neuroscientist!

Taste

Let's finish this lecture by briefly touching on taste. As mentioned, most of what we think of as taste is actually smell, as the ORNs routinely get activated by the food we eat. That also explains why orange juice, for instance, didn't have much taste when I had COVID-19. But in addition to detecting flavors in food through smell, there is also actual taste, the nonolfactory component of the sensation you experience when you put something in your mouth.

As a sensory system, taste is closely related to olfaction, following the same basic principles but with some peculiarities. One difference is the limited range of sensations in taste compared with olfaction: you may think that food has a large variety of flavors, but, again, it's mostly smell, not taste. Although there are many chemicals in our food, our brain recognizes only a limited set of tastes, the *tastands*, and all for a good evolutionary reason. In fact, we can "taste" only five things: (1) sugar, to identify food rich in carbohydrate calories; (2) salt, to detect sodium, an essential ion; (3) umami, to find out if the food is rich in proteins; (4) sour, the taste of acid, to detect if food is spoiled; and (5) bitter-tasting alkaloids, to detect possibly poisonous food.

Another peculiarity is that taste is essentially hardwired, at least in mice: their behavioral responses to these tastes are mostly innate, genetically programmed, in contrast to smell, because most smells acquire meaning through experience (i.e., you associate a certain smell with a food, person, etc.). But we do have some acquired tastes; through experience, you may learn to like bitter or sour flavors, like coffee or beer, which initially you probably disliked. And like all senses, including smell, taste sensations are ultimately constructs of the brain.

Gustatory Pathway

The brain recognizes taste through a gustatory pathway that starts in taste buds, located in the tongue and pharynx. There we have taste cells with taste receptors, which project to the gustatory nucleus in the brainstem, which then projects to the thalamus and, finally, to the gustatory cortex (figure 8.11).

Let's dive in, one step at a time. First comes the tongue. If you look carefully at your tongue, you will see that you have little holes and dots (figure 8.12). Interesting, no? I bet many of you never noticed this before. They are pores and papillae, tiny mushroom-looking structures in whose grooves (which concentrate tastands) we find taste buds (figure 8.13). Taste buds are onion-looking capsules in the epithelium of the tongue. They contain taste cells with an apical surface that contacts the mouth cavity and where taste receptors are located, just like ORNs had receptors in their cilia. Taste cells have different types of taste receptors (figure 8.14): some are ion channels, such as amiloride-sensitive Na^+ channels responsible for detecting salt, or H^+-sensitive proton channels, which detect sour substances. As Na^+ and H^+ are cations, they directly depolarize the

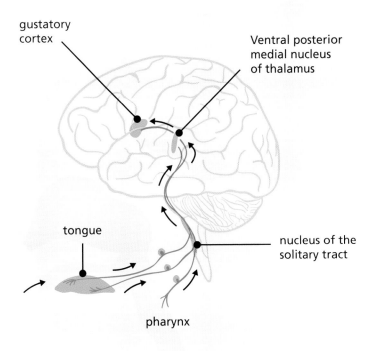

gustatory
cortex

Ventral posterior
medial nucleus
of thalamus

tongue

nucleus of the
solitary tract

pharynx

8.11 The **gustatory system** starts with the transduction of tastands in the tongue by sensory neurons, which send this information to the brainstem, thalamus, and gustatory cortex.

neurons, so no second messenger is needed (figure 8.15). Done! But the detection of sweet, umami, and bitter tastands is more complicated; we can detect them thanks to G-protein-coupled receptors, as in olfaction. These GPCRs activate cation channels, members of the transient receptor potential (TRP) channel family, which depolarize the cells. The end result is the transduction of a chemical stimulus into an electrical signal.

In the basal region of taste cells, the depolarization generated by the channels triggers transmitter release, which generates action potentials in gustatory sensory neurons, whose axons relay that signal to the

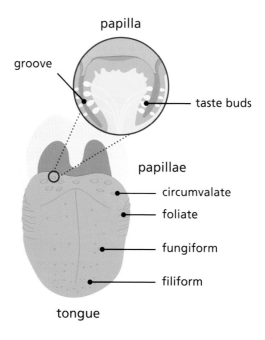

8.12 The **tongue** has different types of papillae with grooves filled with taste buds.

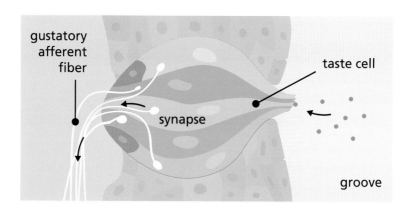

8.13 **Taste buds** have taste cells that detect tastands and generate electrical signals that are sent to the brain.

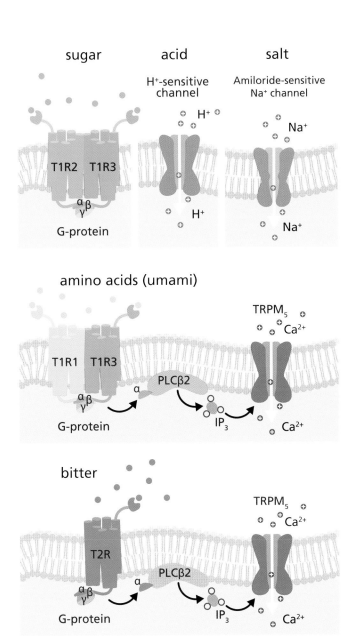

8.14 Different types of **taste receptors** detect different tastands and depolarize taste cells.

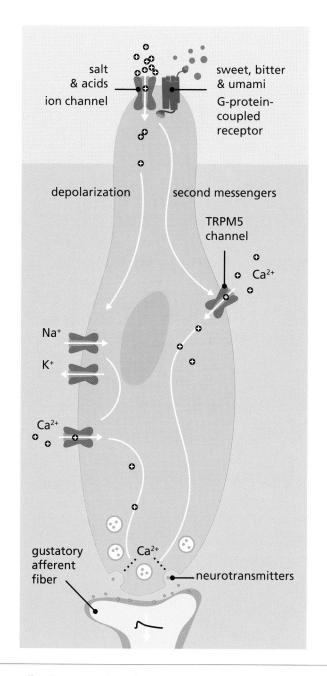

8.15 Taste cells release transmitters that activate gustatory nerve fibers.

nucleus of the gustatory tract in the brainstem (figure 8.11). From the brainstem, taste information is relayed to the hypothalamus, amygdala, and ventro-posterior medial (VPM) nucleus of the thalamus, which projects to the gustatory area in the insular cortex of the frontal lobe, where the five tastes are mapped into five different patches. These anatomical projections make sense: the brainstem stopover, as is de rigueur for most senses, probably mediates head-based reflexes (like when you spit out bad food or throw up), whereas the link to the hypothalamus and amygdala likely helps connect tastes with emotional responses. Finally, the thalamic stop could help us pay attention to particular tastes, and the cortical projection could help us associate taste with particular food and also integrate taste into our cognition.

The relative simplicity of the gustatory system makes it one of the best-understood sensory systems. The fact that it is hardwired and limited to five taste pathways, each originating in taste cells expressing only one type of taste receptors, has enabled elegant genetic manipulation of the gustatory system in mice. By knocking out or knocking in receptors for particular tastands, researchers can actually change the mouse's behavior toward food. For example, wild-type mice lick strongly in response to sugar (no surprise there, as you would probably do the same). But if you knock out sweet receptors, mice act as if they cannot taste the sweetness. Moreover, if you express a bitter receptor in a neuron that is supposed to respond to sweet, the mouse suddenly starts liking bitter substances. The same happens if you activate corresponding brain regions. Stimulating the bitter area of the cortex while the mouse was licking water makes the mouse decrease its licking, as if it were tasting something bitter. The reverse was true if the sweet area of the cortex was stimulated.

Findings from these experiments involving the gustatory system are consistent with a labeled-line code, in which individual neurons respond only to one tastand, and the taste quality is then determined via an anatomical pathway, which is hardwired into behavior. In other words, the identity of the taste is determined by the origin of the axons. If they arise from a sweet taste cell, the taste will be sweet, and if it starts in a bitter taste cell, it will be bitter. These experiments also show that the basic attractive behavioral responses to sweet and repulsive ones to bitter are hardwired—not just the sensation of taste but also the accompanied behavior.

▼ RECAP

The chemical senses—olfaction and taste—nicely exemplify the general principles of sensory processing: they have been optimized (in some cases) to the physical limit (single-molecule detection); they are built with parallel processing (labeled lines in the case of the gustatory system); they track the identity of the stimuli in a chemical space (i.e., chemical-receptive fields); and they relay these signals through a hierarchical set of brain nuclei to build topographic maps of stimulus space in the cortex. The brain also exercises descending control on these sensory pathways, and that probably shapes and actively creates our perception of smells.

The encoding of olfactory and gustatory information via specific receptor molecules has enabled the genetic manipulation of the system for a deep dive into two basic computational principles. For olfaction, this general strategy appears "softwired" based on a neural network strategy, with, first, encoding of the stimulus in a multidimensional space (with as many dimensions as olfactory receptors), randomly scrambling it all, and then decoding this combinatorial code with perceptron-type circuits and attractors that identify particular odors. The gustatory system appears simpler and hardwired: five parallel labeled lines independently carry tastand information to the brain.

Further Reading

Axel, R. 2005. "Scent and Sensibility: A Molecular Logic of Olfactory Perception (Nobel Lecture)." *Angewandte Chemie International Edition* 44, no. 38: 61102–107. An illustration of the power of molecular biology when applied to neuroscience, focusing on the spatial coding of information.

Laurent, G. 2002. "Olfactory Network Dynamics and the Coding of Multidimensional Signals." *Nature Reviews Neuroscience* 3: 884–95. A thoughtful discussion of the role of temporal coding in the olfactory system.

Roper, S. D., and N. Chaudhari. 2017. "Taste Buds: Cells, Signals and Synapses." *Nature Reviews Neuroscience* 18, no. 8: 485–97. A review of the first step in the taste pathway.

Yarmolinsky, D. A., C. S. Zuker, and N. J. Ryba. "Common Sense About Taste: From Mammals to Insects." *Cell* 139, no. 2: 234–44. A comprehensive review of the labeled line design principle of the gustatory system.

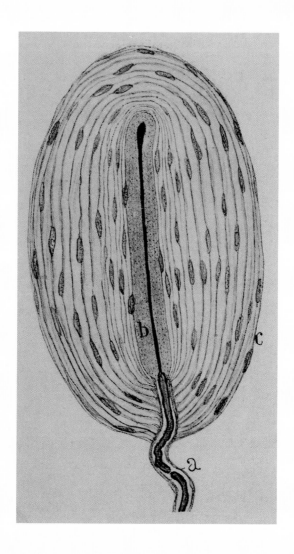

Human Pacini corpuscle. These onion-like structures in our skin are specifically designed to detect vibrations. Courtesy of the Cajal Institute, "Cajal Legacy," Spanish National Research Council (CSIC), Madrid, Spain.

LECTURE 9: TOUCH

During the year 1889 my activity continued with unabated vigor and diligence, applying myself to various neurological problems, but concentrating particularly upon the study of the spinal cord in birds and mammals. In attacking this question, whose obscurity I knew well from experience, explaining the organization of the spinal axis in my anatomical lectures, I was moved first of all by the goal to elucidate as far as possible the difficult problem of the ending of the posterior or sensory roots.

—Santiago Ramón y Cajal, *Recollections of My Life* (1917)

OVERVIEW

In this lecture, we'll learn

- ▶ How mechanical stimuli are transduced into electrical signals in our skin
- ▶ How the spinal cord processes different types of touch information
- ▶ How the cortex contains somatosensory maps of the body
- ▶ How cortical maps change depending on their use

The Somatosensory System

The sense of touch is the most direct way we interact with our environment; we use physical contact to extract information about the position, shape, surface, movement, consistency, and temperature of objects. Evolution has designed a system to take all these measurements, called the *somatosensory system* (from the Greek *soma* = "body"), and it has naturally placed it on the skin, with an inner outpost of it (the *proprioceptive system*, which we will leave for later) inside our body to monitor our limbs, joints, and muscles.

Like other sensory systems, the somatosensory system has many parallel channels, and, once again, acting as a good engineer, the somatosensory system carefully measures amplitude, duration, and phase of the stimulus. Using that information it builds a representation of the object, one that, as noted before, is constructed by our internal model of the object and our memories of past experiences. And, as we will learn, three of the principles of neuroscience—hierarchical processing, mapping, and plasticity—are as crystal clear here as anywhere else.

In fact, there are actually two somatosensory systems, each with a different anatomy and function (figure 9.1). They run parallel to each other from the skin to the spinal cord, to the thalamus and end in the somatosensory cortex. The main one, also known as the *dorsal column system* because it runs along the dorsal white matter of the spinal cord, is evolutionarily modern and uses thick, fast transmitting myelinated axons. Remember myelin, that membrane insulator generated by oligodendrocytes that we discussed in lecture 3? Here we will put that knowledge to good use. This dorsal column system has many subchannels that carry information about touch, vibration, pressure, and proprioception (i.e., the monitoring of the position, state, and movement of our muscles and bones; see table 9.1).

Meanwhile, the secondary pathway, the *anterolateral system*, runs along the anterolateral white matter of the spinal cord. This system is evolutionarily older; has thinner, slower-transmitting axons; and carries nociceptive pain, temperature, and coarse or sensual touch. As pain processing has interesting peculiarities, we will cover it in the next lecture.

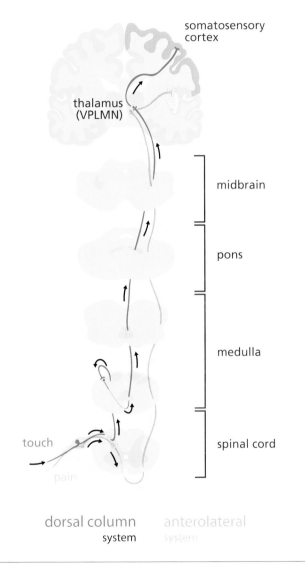

somatosensory
cortex

thalamus
(VPLMN)

midbrain

pons

medulla

spinal cord

touch

pain

dorsal column
system

anterolateral
system

9.1 Touch information is channeled by the **dorsal column pathway**, which starts in the skin and ascends through the spinal cord to the medulla, thalamus, and somatosensory cortex.

TABLE 9.1 The Somatosensory Pathways

Pathway	Dorsal column	Anterolateral
Axons	Thick, myelin	Thin, no myelin
Speed	Fast	Slow
Evolutionary origin	New	Old
Function	Touch	Pain
	Vibration pressure	Temperature
	Proprioception	coarse touch
		Sensual touch

Skin: Channels and Receptor Cells

Let's start with the dorsal pathway. As always, we know much more about the structure than about the function of the brain. As we did with other sensory systems, and given that, So let's follow the anatomy and walk our way from the periphery to the cortex, peppering this march with comments about the receptive field properties and potential functions of the neurons we encounter.

The dorsal pathway starts in the skin, itself a fascinating structure with an impermeable epidermis for protection and to save our precious water from evaporating, and a supportive dermis below the epidermis, which brings in blood and supplies (figure 9.2). Down in the dermis is where we find the majority of nerve endings, in receptors specialized for different modalities of touch. These nerve terminals are loaded with mechanosensory channels that transduce pressure or vibration into electrical signals.

There are three basic types of mechanosensory channel (figure 9.3): a first group (figure 9.3, top) that literally stretches open as a result of membrane tension; a second one (figure 9.3, middle) that has an extracellular anchor (think tip links), which is elastic and attached to a "lid" that keeps the channel closed; and a group (figure 9.3, bottom) in which the elastic spring is connected to a membrane protein that turns on a second-messenger pathway that opens the channel. All three channel types flux sodium and other cations, thus depolarizing the cell. The first two types are fast, as in auditory transduction, whereas the third type, as in phototransduction and olfaction, has slower kinetics but it can amplify the signal due to the intermediate biochemical steps.

These channels are located in touch receptors in the skin, which are quite interesting and come in many different flavors, each measuring a particular aspect of the sensory stimulus (figure 9.2). In fact, the sense of touch can be thought of as several different senses of touch, mediating all kinds of mechanical information provided by these parallel systems acting in concert.

One way to classify touch receptors is by the temporal properties of their responses, either *slowly adapting* (or SA), which means that they respond to a steady skin indentation with sustained electrical discharges, or *rapidly adapting* (RA) receptors, which stop firing as soon as the indentation becomes stationary (figure 9.4). Because of these properties, in SA receptors, the spiking frequency is directly proportional to the total amount of pressure; that is, proportional to the force of the indentation.

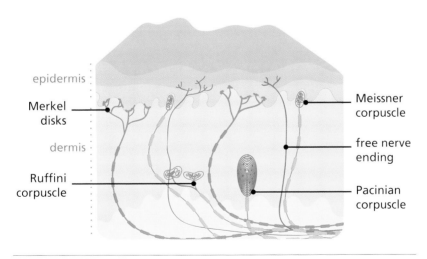

9.2 The skin has different types of **somatosensory terminals** that detect different touch modalities.

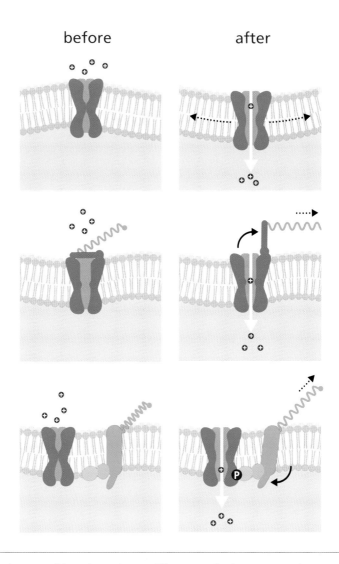

9.3 Mechanosensitive channels use different mechanisms to transduce touch into electrical potentials.

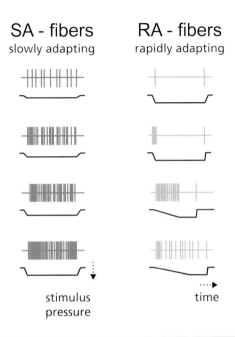

SA - fibers
slowly adapting

RA - fibers
rapidly adapting

stimulus
pressure

time

9.4 Slowly and rapidly adapting receptors specifically detect the amplitude or phase of the mechanical stimulus.

These SA receptors are better suited to monitoring the strength, duration, and shape of the stimulus.

Meanwhile, RA receptors stop firing if a stimulus persists, and therefore, they detect temporal onsets and offsets of discrete stimuli, such as changes in the temporal and spatial patterns of stimulation. Since we have learned about audition, this strategy should be no news to you: nature is again measuring the amplitude and the phase of stimuli, this time with parallel receptors, one set (SA) specialized for amplitude and another (RA) for phase. It gets even better. It turns out that there are at least four major subtypes of SA and RA touch receptors in the skin: Meissner corpuscles (RA) and Merkel disks (SA) near the surface, and Pacinian (RA) and Ruffini (SA) corpuscles deeper in the dermis. Each subtype has a characteristic morphology, a specific location in the skin, receptive fields, and responses to touch (figures 9.2 and 9.5).

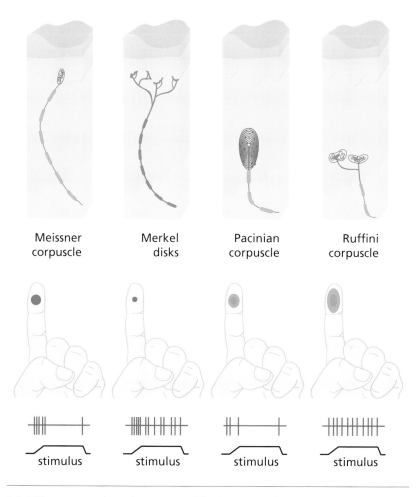

Meissner   Merkel   Pacinian   Ruffini
corpuscle   disks   corpuscle   corpuscle

9.5 Different types of **touch receptors** differ in structure, location, receptive fields, and responses.

As an example, let's look more closely at how one of these four sub-types of receptors, Pacinian corpuscles, work. They are particularly striking and beautiful (see Cajal's drawing at the opening of this lecture) and demonstrate the logic behind their design. Pacinian receptors are rapidly adapting and have an axon terminal enveloped in onion-like layers of fluid-filled connective tissue that works to dampen the applied pressure

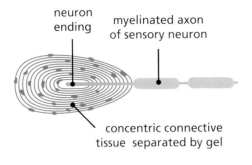

neuron
ending

myelinated axon
of sensory neuron

concentric connective
tissue separated by gel

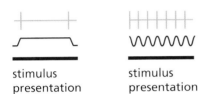

stimulus
presentation

stimulus
presentation

9.6 Pacinian corpuscles have layers of tissue surrounding the nerve ending, which provide them with rapidly adapting properties.

(figure 9.6). If the receptor didn't have this onion-looking structure, it would respond to the stimulus all the time, as long as the stimulus was pressing on the nerve ending (figure 9.7). But because the fluid-filled layers absorb and redistribute the pressure, the nerve ending is protected from mechanical distortion, as if by a mechanical shock absorber. Because of that onion-like protection, action potentials are only triggered when the stimulus is first applied, when the initial pressure wave propagates and stretches the membrane, thus opening the channel.

Pacinian corpuscles also respond at the end, when the stimulus is removed, because a negative pressure wave propagates and also stretches the membrane the other way. Mathematically, this means that a Pacinian corpuscle receptor uses its mechanical properties to take the derivative of the stimulus; that is, the rate of change of pressure being applied. Amazing, no? So when we look at Cajal's beautiful drawing of a Pacinian corpuscle, we are actually looking at nature's mechanical derivative machine.

But why bother taking the derivative? Simple: to measure vibration, you need to quickly compute the changes in the touch stimulus as a function

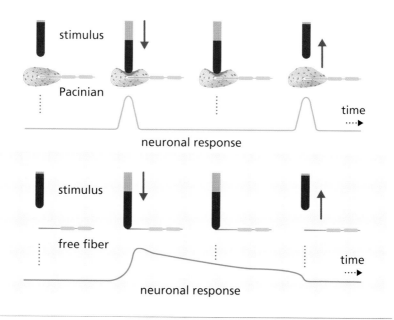

9.7 Removing the onion-like **envelope** of a Pacinian corpuscle turns it into a slowly adapting receptor.

of time. And what is the big deal with vibrations? Why do we need to measure them? Vibrations actually tell you about the structure beneath the surface of an object; it's a way to "see" deeper, to feel the mechanical properties of the object. Also, when you touch something you move your fingers across an object, and this tells you about the texture of its surface, transforming it into vibrations as well.

Finally, how good are these Pacinian receptors? They can respond to vibration frequencies as high as 500 Hz, which means firing impulses every 2 ms. Not bad! And all of that comes from a beautiful match of structure to function. Don't you just love biology?

And what about the other receptors? We don't know for sure, but following the nice story of how Pacinian receptors work, keep in mind that each of the different types of receptor are likely just as sophisticated and precise. All together they likely do a killer job measuring and mapping the somatosensory inputs and sending that information up to the brain in parallel channels. And for each of these different subsystems, the brain also keeps track of the intensity of the stimulus by the number of active neurons in each receptor population, and this population code depends on the intrinsic sensitivity of each neuron to a stimulus. Since neurons differ in sensitivity, a softer stimulus activates only low-threshold neurons, followed by high-threshold neurons as stimulation intensifies. As with other sensory systems, the dorsal pathway does a complete, ordered, and systematic job measuring the touch stimuli and turning them into a parallel set of electrical signals that it sends to the brain.

Dorsal Root Ganglia

Let's climb up the pathway. What happens to all this carefully collected information as it leaves the skin? The axons of the nerve terminals from the skin, muscles, and joints of the limbs and trunk are generated by neurons that are clustered together in dorsal root ganglia (DRGs), located immediately adjacent to the spinal cord. DRG neurons are part of the peripheral nervous system, as you recall from lecture 2, and are pseudo-unipolar in shape. That means they have a T shape with two branches that stem from the same cell body branch. The peripheral branch terminates on the skin and muscles as a free nerve ending or on specialized receptors. Don't let the nomenclature confuse you: the peripheral branch is called an axon, but it actually is more like a dendrite, receiving sensory inputs. The central process, the proper axon, enters the spinal cord and either terminates within the gray matter—forming reflex circuits, which we will study later—or ascends to nuclei in the medulla and the rest of the brain. DRGs are located at every level of the spinal cord and are split into cervical (eight vertebrae), thoracic (twelve vertebrae), lumbar (five vertebrae), and sacral (five vertebrae) segments.

Imagine now that you insert an electrode into your hand and record from the axons that come from that spot (figure 9.8). You will discover

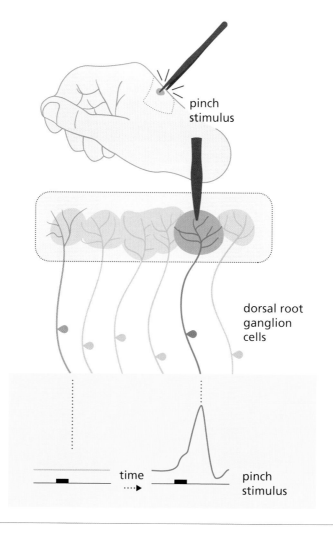

9.8 The **receptive field** of a DRG neuron is the skin area that makes it fire when you touch it.

that action potentials occur only if you touch a particular patch of skin. As you may remember, the skin area, or region of space, in which a stimulus can activate a sensory neuron is called the neuron's receptive field. This is one of the most important concepts of neuroscience; it was actually discovered by Sherrington in the somatosensory system but applies to all sensory systems. The dimensions of the receptive fields vary, with small receptive fields in more sensitive skin areas like our fingertips and large ones in our back, for example. Thus, the size of the stimulus determines the total number of neurons activated, and the neural representation of an object is a collective mosaic of individual receptive fields. So you just discovered receptive fields. If you had conducted that experiment in 1900, ahead of Sherrington, you would have received a Nobel Prize!

Now let's continue this thought experiment. If you are careful to measure receptive fields from different axons and annotate how long they take to generate an action potential after you activate the skin, you will find out that some axons are fast, generating action potentials shortly after you touch the skin, whereas others are slow. And if you were to look through your microscope at those axons, you would find that fast axons are thicker, whereas slow axons are thinner.

As you may recall from the discussion of electrical properties of axons and cables in lecture 3, a mathematical relation states that current flowing through a cable is directly proportional to the width of the cable and indirectly proportional to the membrane's resistance, since current doesn't flow easily through the plasma membranes because they are lipids. Guess what? The fast axons are not only thicker, they are also heavily myelinated, whereas the slow axons are thin and unmyelinated. Myelin acts to insulate the axon by decreasing the denominator. In reality, the somatosensory system has different classes of axons, classified according to their speed (figure 9.9). The thickest and fastest innervate skeletal muscle and carry proprioceptive information coming from inside your body. This makes sense, as you need to be able to quickly monitor your muscles, tendons, and skeleton during movement. The next-fastest axons innervate the skin mechanoreceptors. The slowest axons are thinner and innervate nociceptors and thermoreceptors, and carry pain and temperature information, and also coarse touch and itch via the anterolateral pathway. Because of this, messages from the dorsal pathway arrive at the CNS faster than those from the anterolateral pathway.

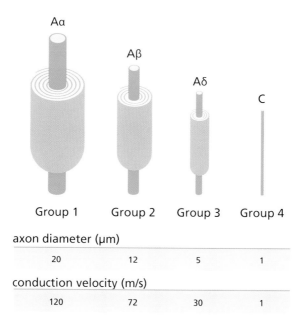

axon diameter (µm)			
20	12	5	1
conduction velocity (m/s)			
120	72	30	1

9.9 Somatosensory axons propagate action potentials at different speeds, depending on their width and myelin wrapping.

Spinal Cord

As we follow these axons into the CNS, we arrive at the spinal cord. This complex yet delicate structure is one of the marvels of neuroscience, as Cajal's opening quote indicates. It mediates the relationship between our physical environment and the brain, bringing in somatosensory information and controlling most of our muscles. To understand the spinal cord's structure, let's go back to neural development, something I always love doing, following Cajal's dictum. In fact, in his 1917 autobiography, he argued that the key to his success was his study of tissue from young animals and embryos, because at that stage, axons don't have myelin and you can clearly see the neurons and their patterns. He said, "Why not study the young forest, as one would say, while it is still in the nursery stage?"

Well, as typical bilaterians, humans are segmented animals, which means that our bodies develop from many bilaterally paired repeating blocks (or lumps) of mesoderm called *somites*. What is the mesoderm? It is all the stuff between the ectoderm (future skin and nervous system) and endoderm (future organs and guts). In a human embryo, the first somite appears early during pregnancy, and after that, more somites develop continuously and quickly along both sides of the neural tube (the precursor of the central nervous system). Somites are transient structures that themselves later differentiate into dermatomes, myotomes, and sclerotomes, precursors of the deep layer of the skin (dermis), the muscle, and the skeleton and connective tissue, respectively. Then, as neurogenesis begins, the nervous system develops nerves that innervate all of the tissues derived from that particular somite.

An effective system of segmental units thus develops, in which sensory and motor information from a specific segment of the skin (a *dermatome*; figure 9.10) travels through a dedicated spinal nerve and is fed to different regions of the butterfly-shaped spinal cord. The nerve bifurcates into the dorsal root ganglion for afferent sensory signals and the ventral root for efferent motor signals. Because of this ordered segmental mapping of dermatomes and spinal nerves and segments, dermatome maps provide an important diagnostic tool for locating injury in the spinal cord and dorsal roots.

The dorsal and anterolateral pathways differ in their anatomy (figure 9.11). As they enter the spinal cord, the axons from the dorsal (touch) pathway (which come from the so-called first-order DRG neurons) either make synapses on neurons in the dorsal horn of the spinal cord, creating a reflex arc (more about that later), or continue upward through the dorsal column of the spinal cord all the way to the medulla, where they synapse on a second-order neuron. Axons from these second neurons then decussate (i.e., cross the midline) to the other side of the brain and proceed to the thalamus, where they synapse on third-order neurons. These third and final neurons in the pathway extend their axons into the cortex (figure 9.12). The overall result is that sensory information from one side of the body ends up being processed in the opposite part of the cortex.

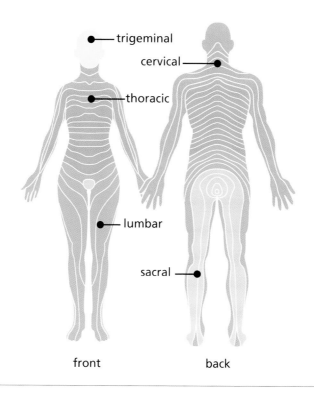

front   back

9.10 Our skin is divided into **dermatomes**, areas innervated by a particular DRG.

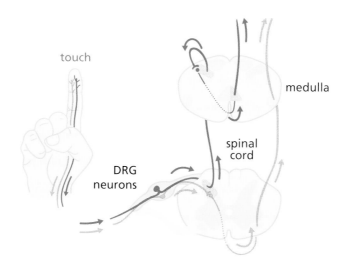

9.11 DRG axons from the touch pathway contact neurons in the dorsal horn of the spinal cord or in the medulla.

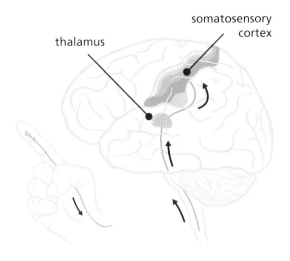

thalamus

somatosensory cortex

9.12 The **touch pathway** goes from the skin to the thalamus and somatosensory cortex.

Meanwhile, in the anterolateral (pain) pathway, the first-order neuron terminates in the dorsal horn of the spinal cord, synapsing on a second neuron, whose axons decussate straight from there and travel all the way to the thalamus through the anterolateral column of the opposite side of the spinal cord. The third neuron again receives the input from the thalamus and delivers it to the cortex, where it meets the axons from the dorsal pathway.

To reiterate, each pathway consists of three neurons, and the main difference between them is simply the point of decussation—crossing to the other side of the CNS. Whereas the dorsal pathway decussates in the midbrain, the anterolateral pathway decussates in the spinal cord, but at the end, they both end on the opposite side of the brain, as occurs with other sensory systems.

Somatosensory Cortex

What happens to the sensory information once it reaches your brain? How do you know that you pinched your finger or that a friend touched your shoulder? Let's dwell for a second on the medullar and thalamic stops along the dorsal pathway—think parallel pathways, as in vision or audition. Each of the receptor types in the skin extracts a type of information from the stimulus, and this starts a pathway that is fed side by side with the other pathways. How many parallel somatosensory channels? The last lecture I heard on this topic suggested that there are about a dozen of them, some slowly adapting and others rapidly adapting. And all this information is organized in somatotopic maps (i.e., maps of the body), with a precise correspondence to a point on the body and its representation in the brain, structured dermatome by dermatome. Some basic processing of this information takes place in the medulla, and farther up we encounter the usual mystery about what the thalamus does—maybe it uses attention to select sensory information? We just don't know.

But we do know a few things about the somatosensory cortex. Somatosensory information from the thalamus arrives to the primary somatosensory cortex in the postcentral gyrus of the parietal lobe (figure 9.13). Specifically, the primary somatosensory cortex (S1) has several subregions: areas 3a, 3b, 1, and 2. The bulk of thalamic neurons project to areas 3a and 3b, which project to areas 1 and 2. Areas 3b and 1 receive their inputs from receptors in the skin, whereas 3a and 2 receive proprioceptive information from receptors in the muscles, joints, and skin. Thus, neurons in S1 are three steps removed from the stimulus (DRGs–medulla–thalamus–cortex).

Each cortical neuron ends up receiving information from a specific class of receptors in a specific area of the skin and it has a have larger receptive field than that of the original receptor. We perceive that a particular region of the skin has been touched because of the neuronal response in the cortex. But the reverse can occur: you can hijack the pathway and generate a hallucinatory experience of touch by directly stimulating a specific area of the cortex. In fact, you will feel the same sensation whether it's from a real stimulus or when your cortex is stimulated. So one could argue that the conscious perception of somatosensation occurs in the cortex; it's created by the cortex using information coming from

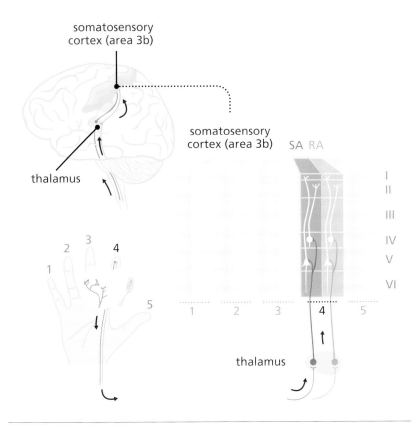

9.13 The **primary somatosensory cortex** has a columnar organization or receptive fields.

the periphery. Again, this demonstrates the same Kantian theme as in the other sensory systems.

Cortical Maps

The primary somatosensory cortex is neatly organized in maps. Surprise, surprise! These somatosensory maps are built according to a stimulus feature in a complex yet extremely logical structure. Let's start from the beginning.

You may remember that the cortex—or, to be more specific, the neo-cortex (the larger and evolutionarily more recent part of the cerebral cortex, as opposed to the archicortex-hippocampus and the paleocortex-entorhinal cortex)—has six layers that run from the outer surface to the inner white matter. The thickness of these individual layers and the details of their organization vary throughout the cortex, as different parts of the cortex are specialized to receive specific types of information (primary motor cortex, primary visual cortex, prefrontal association cortex, etc.). I mention this because in 1957, ahead of Hubel and Wiesel, Vernon Mountcastle discovered that neurons in the primary somatosensory cortex have similar receptive fields as you go down in depth. He is the one who came up with the name *column*. He found that S1 is arranged in vertical columns or slabs, traversing all six cortical layers (figure 9.13). These columns receive input from a specific body part, so that together, they form a somatotopic map. And each column is divided into smaller ones, which receive inputs from SA and RA receptors separately. While receptive fields of the DRG SA1 and RA1 fibers are tiny spots on the surface of the hand, the cortical neurons receiving their information cover the entire fingertip or several adjacent fingers. Receptive fields in higher cortical areas cover an even larger area—spanning whole functional regions of skin.

Let's talk more about these maps. Just as dermatomes are connected to specific nerves, creating a map of the body, the somatosensory cortex has another map of the body (the cortical *homunculus*) embedded in it (figure 9.14). Each area of the body surface is thus represented in the cortex, but there are fascinating differences between our actual body and the brain's version of it in the homunculus. The parts of the body that occupy larger parts of the cortex are not those that are physically larger but those that are more densely innervated by the nervous system. If we were to draw a human according to this map, we would end up with a rather ugly creature with huge lips and enormous hands (figure 9.15). This is actually what our brain cares about, and it is a fascinating insight into who we are. If you were an extraterrestrial looking at this map, you would conclude that humans care a lot about their mouths (speech) and their hands (dexterity). Talking monkeys that do things with their hands. As Heidegger said, humans are toolmakers.

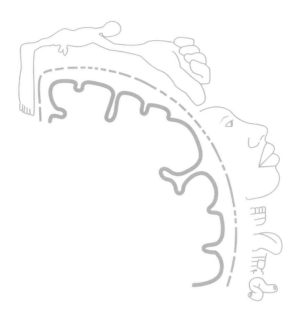

9.14 The primary somatosensory cortex has a map of the body, the "**homunculus**."

9.15 The cortical homunculus has an **expanded representation** of hands and mouth.

Cortical Plasticity

Something important happens with these maps. Given how hard it is to wire things up during development, you may think that sensory maps in the cortex are set for life. But, guess what, they change depending on experience. In our cortical homunculus, the size of different body regions enlarges or shrinks depending on how much we use them. We know this from experiments mapping the homunculus in monkeys, in which you can observe dramatic changes in cortical maps due to use or disuse. For instance, if you train a monkey to repeatedly use the tips of a finger for one hour a day, after that training, the cortical representation of that finger increases (figure 9.16). I'll bet you that because of smartphones, the representation of our thumb areas has grown a lot!

Similar experiments also suggest that the somatosensory cortex is continuously adapting to the outside world. In fact, referring back to the beginning of the book, one could argue that these maps—these cortical representations—*are* the world, your world, what you care about. So you could argue that the function of the cortex is to build a mental representation of the world, a representation that you then manipulate internally. For example, if you imagine that you are being touched without any actual input to your skin, your cortical homunculus is

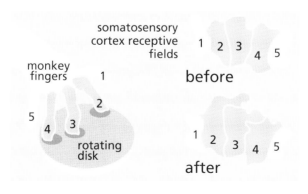

9.16 Repeated use of particular fingers increases their **cortical representations**.

activated. The cortex can be externally or internally driven. It all starts to make sense.

Higher Somatosensory Cortex

Moving upward, we see that neurons from the primary somatosensory area then project to higher centers in the cortex: the secondary somatosensory cortex (S2), posterior parietal cortex, and primary motor cortex (figure 9.17). So what happens beyond S1, in S2 or the posterior parietal cortex? What types of representations of the world are mapped there? We don't quite understand what is going on, but it is clear that as information moves into higher cortical areas, the receptive fields of neurons keep increasing in size. Just like DRGs, SA and RA receptive fields are small

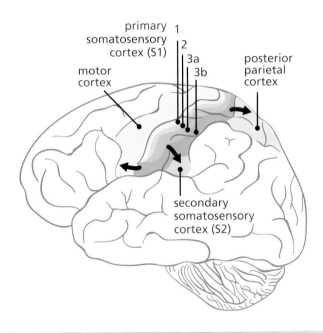

9.17 The **primary somatosensory cortex** receives thalamic touch information and relays it to motor and parietal cortex.

dots on the hand, and receptive fields of primary somatosensory cortical neurons cover the entire fingertip or several adjacent fingers, receptive fields in higher cortical areas cover an even larger area—spanning whole functional regions of skin. It seems that you are constructing mental images of your body in a piecemeal fashion.

In fact, these secondary somatosensory areas are just the beginning of a hierarchy of cortical areas with increasingly more sophisticated receptive fields and processing. Cortical areas involved in the early stages of sensory processing are concerned with only one primary sensory modality, so we refer to them as *unimodal association areas*. Information from unimodal association areas then converges on multimodal association areas, which combine sensory modalities. It is as if they gather information about the world from separate sources and then put it all together to create a more complete picture of the world. The pathway represents an increasingly more sophisticated understanding of the world. For example, the primary S1 simply alerts the brain that you are touching an object, while the secondary cortex interprets the different inputs to conclude that the object you are touching is round or square, big or small, hard or soft or pliant.

Descending Control

To finish our discussion of touch, it turns out that, like in all sensory systems, the cortex not only represents the world but likely actually constructs this representation by influencing what we perceive. Not all information processing occurs in a bottom-up manner. The cortex can also affect our perceptions of touch through top-down processes. In fact, if you count axons going up and down, in some sensory systems, there are ten times more axons going down than up. Why so much top-down control? Attention is one example of a top-down process. If you anticipate being touched on your hand, the mere fact of paying attention to the hand will cause a greater activation of the higher cortical areas once the touch actually occurs. More generally, you could argue, going back to Kant, that the brain is internally selecting, using "categories," which features of the sensory world it cares about in order to build your perception.

A corollary, also following Kant, is that there are features of the sensory world that we do not perceive and therefore our mind has no access to them. As he put it, we are trapped on an island in the middle of the ocean of the real world, limited by our internal categories—by the boundaries of our senses. But maybe science can help us see beyond.

▼ RECAP

The touch system starts in the skin, where specialized receptors transduce all kinds of mechanical information into action potentials, faithfully measuring their phase, amplitude, and frequency, and ship them up through the spinal cord, brainstem, and thalamus, until they reach the somatosensory cortex. There, the brain builds careful topographic maps of the world, keeping all these different types of information neatly organized in separate regions, and uses these maps as a mental representation of the body and, in higher cortical areas, of the physical world. Finally, descending pathways from the cortex select and control all upcoming information. The somatosensory system nicely recapitulates the principles of sensory processing: parallel processing (labeled lines), measurements of phase and amplitude, receptive fields, topographic maps, hierarchical processing for abstraction of stimulus properties, descending control and construction of sensory percepts, and experience-based plasticity. The somatosensory system is indeed where many of these basic principles of neuroscience were discovered. And all of this is done probably to generate a mental model of our body as part of a model of the world so we can predict the future and act accordingly.

Further Reading

Mountcastle, V. B. 1998. *Perceptual Neuroscience: The Cerebral Cortex*. Cambridge, MA: Harvard University Press. Mountcastle had a comprehensive and systematic mind and collected everything he knew about the cortex in this volume. This is as good a book about the cortex as it gets.

Mountcastle, V. B. 2005. *The Sensory Hand: Neural Mechanisms in Somatic Sensation*. Cambridge, MA: Harvard University Press. In this monograph, Mountcastle explains how somatosensation works.

Ramón y Cajal, S. (1917) Reprinted 1989. *Recollection of My Life*. Repr. ed. Cambridge, MA: MIT Press. A wonderful autobiographical account of the ups and downs of the life and career of this "science's hand labourer," as Cajal liked to call himself. I cannot recommend this enough.

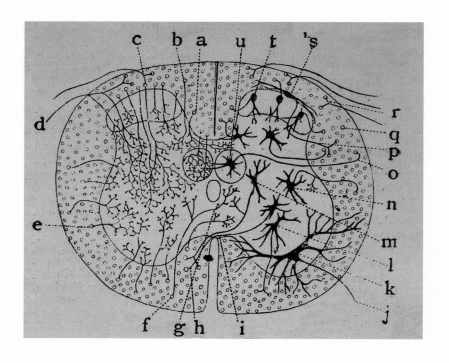

Spinal cord. Complex neural circuits in the spinal cord process pain information. Courtesy of the Cajal Institute, "Cajal Legacy," Spanish National Research Council (CSIC), Madrid, Spain.

LECTURE 10: PAIN

I heard a shout. Starting, and looking half round, I saw the lion just in the act of springing upon me. . . . He caught my shoulder as he sprang, and we both came to the ground below together. Growling horribly close to my ear, he shook me as a terrier does a rat. The shock produced a stupor similar to that which seems to be felt by a mouse after the first shake of the cat. It caused a sort of dreaminess in which there was no sense of pain nor feeling of terror, though [I was] quite conscious of all that was happening. It was like what patients partially under the influence of chloroform describe, who see the operation but feel not the knife.

—David Livingstone, *Missionary Travels and Researches in South Africa* (1857)

OVERVIEW

In this lecture, we'll learn

- ▶ How noxious stimuli are transduced into electrical signals in our skin
- ▶ How specific pathways process noxious information
- ▶ How the feeling of pain is constructed by our brain
- ▶ How the brain can switch off pain

Pain and Nociception

If you have ever had surgery or dental work done, and they used local anesthetics, you know what Dr. Livingstone was talking about. How come you can switch off pain if the lion is still mauling you or the doctor is cutting your skin open? How do we perceive pain? For that matter, What is pain?

First, just as there is a difference between the wavelength and color of an image, there is a difference between a painful (noxious) stimulus and the actual experience of pain. *Nociception*—that is, the brain's detection system for noxious stimuli—generates a painful sensation. Or not. That is the main point of this lecture: nociception and pain are two different things. You can have nociception without pain, as in the case of Livingstone, or you can experience pain without nociception, as in the neurogenic pain syndromes, about which I could talk forever, as a former chronic pain patient.

The feeling of pain is an internally generated response in our spinal cord and brain to a nociceptive stimulus. That explains why the perception of pain, perhaps more than any other sense, is subjective and influenced by many internal or external factors. An identical sensory stimulus can elicit quite distinct responses in the same individual under different conditions. Besides Livingstone's experience, there are lots of examples of this: Many wounded soldiers do not feel pain until they have been removed from the battlefield, and injured athletes are often unaware of pain until the game is over. Simply put, there are no purely "painful" stimuli, sensory stimuli that invariably elicit the perception of pain in all individuals.

The variability of the perception of pain is yet another example of a principle that we have encountered in earlier lectures: pain is not the direct expression of a sensory event but, rather, the product of elaborate processing constructed by the brain from sensory signals. We are back to Kant. Actually, we can go all the way back to the Suttas, where Buddha, without knowing any neuroscience, brilliantly describes pain as two arrows hitting us consecutively. The first arrow is the painful stimulus, which we cannot control. And the second arrow is our reaction to the stimulus, which is much worse but something we can diffuse or disengage. This is an important life lesson, since, as Marcel Pagnol put it in *Le château de ma mère*, the human experience boils down to brief "moments of joy, obliterated by unforgettable sadness."

Skin

As we learned in the previous lecture, nociceptive information is carried by the anterolateral pathway. Remember that the pathway gets its name because after starting in the skin, it climbs via the anterolateral tract of the spinal cord and from there goes to the thalamus, and subsequently to the cortex (figure 10.1). We also discussed the fact that this pathway is evolutionarily older, with thinner axons that typically are not myelinated (or very thinly so), which are therefore slower. But in reality, this anterolateral pathway is made up of many different and independent channels, each with exquisite sensory specificity, more than Buddha ever imagined. There is sharp (first) pain, burning (second) pain, hot, cold and itch sensation, sensual touch, and, finally, also a specific channel to detect lactic acid, a by-product of high metabolism and muscle damage.

Let's march our way up from the skin to the brain, as we did with touch. And just like touch, the pain pathway starts in the epidermis and dermis. In this case, the receptor structures that sense noxious stimuli are the termination of the axons from the dorsal root ganglion (DRGs) cells. These axonal processes take the form of free nerve endings in the skin. Although they all look pretty much the same, there are actually several types of free nerve endings, each specialized for a different job, with different types of channels in the membrane that can be triggered by different nociceptive stimuli.

There are four major types of nociceptive stimuli: mechanical (intense pressure), temperature (very hot or very cold), tissue injury (inflammation), and polymodal (or silent nociception; in our viscera), all of which respond to inflammation and chemical stimuli. Each of the four major types of nociceptive terminals contains different types of receptor channels on the membrane of its nerve endings (figure 10.2). Most of these channels belong to the TRP (transient receptor potential) ion channels family, which, as you will remember from our discussion of taste, are K^+/Na^+ channels. They transduce the noxious stimulus, depolarize the nerve ending, and trigger an action potential in the axon.

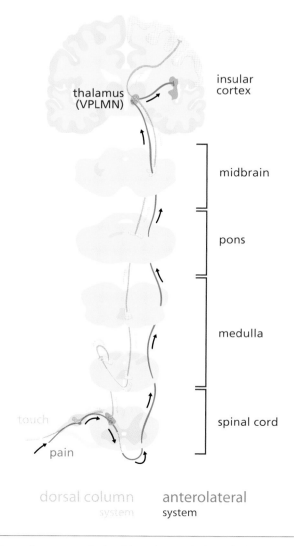

insular
cortex

thalamus
(VPLMN)

midbrain

pons

medulla

spinal cord

touch

pain

dorsal column
system

anterolateral
system

10.1 The **pain pathway** starts in the skin and ascends through the anterolateral tract of the spinal cord, reaching the thalamus and cortex.

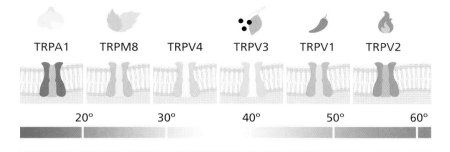

10.2 Different **TRP channels** are activated by specific temperatures and noxious stimuli (from left to right: garlic, menthol, vanilla, pepper, flame).

The variety of temperature-sensitive TRP channels in the pain pathway is thought to underlie the perception of the large range of temperatures that we can detect. So when you touch a cold, warm, or burning object, you are actually activating three specific TRP channels. These channels can also be modulated by particular molecules. One interesting example is capsaicin, the chemical present in peppers, which activates the heat-sensing TRIPV1 channel (figure 10.3). This reaction explains why spicy food feels hot and painful—you are opening the same exact channels that are triggered by hot tem-

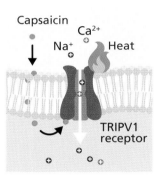

10.3 Capsaicin activates heat-activated **TRPV1** channels, thus mimicking heat.

peratures. A similar story applies for menthol and vanilla, which modulate the activity of cold-sensing channels and may partly explain why mint feels refreshing. Now you understand why we love spicy food or ice cream.

Hyperalgesia

Before we leave the skin, it turns out that, besides carrying nociceptive information to the brain, nerve fibers are involved in reactions to tissue injury. The anterolateral pathway axons, when active, also fire all their branches in the vicinity of the injury, and these axons actually release compounds that trigger an inflammatory response in the nearby skin. These axons thus serve to both receive noxious inputs and also release transmitters locally. So by activating these local axons, you are building a complex local chemical environment.

The substances released by the axons "feed" on each other, and the pain becomes magnified. This is often perceived as the spreading of the pain around the area and is known by a few terms: *hyperalgesia* (the Greek *algesis* for pain; since if you touch the skin next to the area of damage it hurts), *neurogenic inflammation*, or a *pain flare*.

How does hyperalgesia work? Let's say you have a pin stuck in your skin (figure 10.4). Local axons get activated and release a whole slew of compounds, which add to a complex mix of chemicals released from damaged cells that accumulate at the site of the tissue injury. This cocktail contains

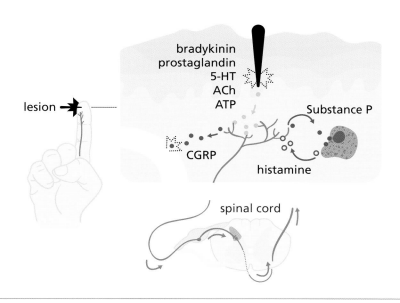

10.4 An **inflammatory response** occurs when the skin is breached and is partly mediated by free nerve endings.

peptides and proteins such as bradykinin, substance P, and nerve growth factor, as well as signaling molecules like ATP, histamine, serotonin, prostaglandins, leukotrienes, and acetylcholine. Those molecules come from different cells and different things. For example, substance P is released by the nerve fibers themselves and causes leakage of plasma to allow macrophages and immune cells to access the damaged tissue. CGRP (calcitonin gene-related peptide) is also released from nerve fibers, and activates nociceptors and causes dilation of the peripheral blood vessels to bring more blood into the area (hence its redness). Histamine is released from local mast cells after tissue injury and activates polymodal nociceptors. ATP, serotonin, and ACh are released from damaged endothelial cells and platelets and together act to indirectly sensitize nociceptors by triggering the release of prostaglandins and bradykinin from peripheral cells. Bradykinin is an active pain-producing agent that directly activates pain fibers and increases the synthesis of prostaglandins. Finally, prostaglandins are released from damaged cells by the activity of the COX (cyclooxygenase) enzyme, which is regulated by many pain meds, such as nonsteroidal anti-inflammatory drugs (NSAIDs).

What is the purpose of all this inflammatory response? Why would nature want to make such a "painful" mess? In reality, all these events point to the same fruitful goal: to establish a battleground so our macrophages and immune system can fight and defeat potential invaders. As the skin is the first line of protection for the body, when it's breached and the defenses are down, you bring in all the troops!

Spinal Cord

Let's now leave the skin behind. If you follow the axons into the spinal cord, you'll notice that they vary in thickness. As mentioned, there are several parallel pathways within the dorsal column system and also within the anterolateral system, and they use different types of axons (figure 10.5). Larger, myelinated axons (known as *A-delta fibers*, to differentiate them from the even thicker A-alpha and A-beta fibers from the dorsal system) carry sharp, fast pain, whereas thinner unmyelinated axons (C fibers)

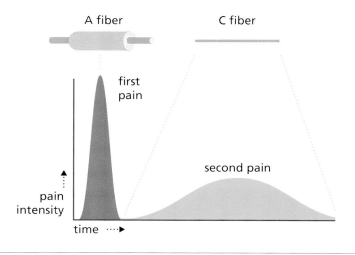

10.5 The **anterolateral pathway axons** have different thickness and insulation, generating fast and slow pain sensations.

carry slower, burning pain and also itch. So even the first arrow of Buddha is split into faster and slower arrows. In fact, the next time you hit your finger with a hammer when trying to nail something, please pay attention and notice three waves of sensation. First, you will feel the hammer hit your finger. This sensation is mediated by the dorsal pathway, with thick, fast axons. Then comes the first wave of pain, mediated by the A-delta fibers. Finally comes the slower, burning sensation, carried by the C fibers. Then comes a series of expletives, which are cortically mediated.

The different anterolateral subpathways remain separated and go to different parts of the spinal cord and the brain (figure 10.6). Remember that the spinal cord has a sensory dorsal horn and a motor ventral horn and is structured in different layers. In this case, the myelinated A-delta synapse onto neurons in the more superficial layers of the dorsal horn, whereas C fibers end up in intermediate layers. Although we don't quite understand why the spinal cord has so many layers and what each layer does, we can still see that information is being sent to specific spots, likely preserving the specificity of the stimulus.

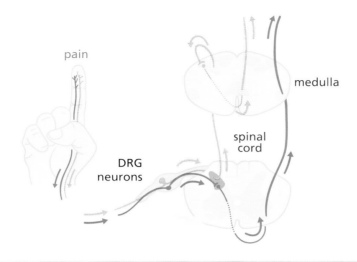

10.6 DRG neurons carrying pain information make synapses in the dorsal horn of the spinal cord, before decussating and ascending through the anterolateral tract.

Importantly, the dorsal horn of the spinal cord is not only full of traditional neurotransmitter (i.e., glutamate) but also peptides, these "poor cousins" of the traditional transmitters that are often ignored. In fact, the same terminals that bring the information from the skin to the spinal cord (the axons of DRG neurons) release both glutamate and peptides like substance P, opiates, and so on. Glutamate is fast acting, and peptides serve to modulate the response to glutamate and make it either stronger or longer. One possibility is that peptides essentially shape the kinetics of the glutamate response, but neuropeptides likely have a much bigger role in brain circuits than we give them credit for today.

Why does the spinal cord contain so much neural circuitry and neurotransmitters? It turns out that peripheral pain can be learned as well as controlled and modulated. If you experience repeated noxious stimuli, the circuitry of your spinal cord becomes potentiated. Now fast forward to the future, and even a small stimulus trigger will produce a large painful response. This process is called *sensitization* and is a form of learning.

You likely have experienced this yourself, but if you haven't, believe me, it's no fun.

How does one "learn" pain? It is actually similar to the way learning happens in other parts of the brain. If you stimulate the C fiber repeatedly, the response will become much bigger, because synapses become more efficient (potentiated). Just as in your typical forebrain neuron, long-term potentiation in the nociceptive pathways occurs by NMDA receptors, which mediate the influx of calcium at the synapse. Moreover, just like in the hippocampus and cortex, neurotrophins (like NGF and BDNF) are also implicated.

You may recall from the development lecture that neurotrophins are a family of proteins that induce survival, development, and function of neurons. Here they play a different role, serving as molecular signals for the synapse to get stronger. The way this works is that neurotrophins are released by the postsynaptic neuron and bind to TrKA receptors on primary nociceptors, triggering localized post-translational changes in expression of ion channels that increase nociceptor excitability.

One last question before we move on. Why would the spinal cord want to learn pain? I guess so you can avoid the stimulus at all costs. They say fear and pain are one-shot teachers. You never forget that sticking your finger in a light socket is not a good idea. We will see this again when we discuss the amygdala in lecture 15, on emotions.

Spinal Lesions

Let's go back to the decussations, the place where a pathway crosses the midline, switching to the other side of the nervous system. Decussations differ for the touch and pain systems. For the anterolateral pathway, this happens at the spinal cord level when the axons from the DRGs come in (figure 10.6), whereas for the dorsal pathway, the decussation occurs in the medulla. Because of this, pain information crosses over at the dorsal horn, whereas touch information crosses in the medulla. Why is that interesting? You can use this information to identify the precise location of a neurological lesion.

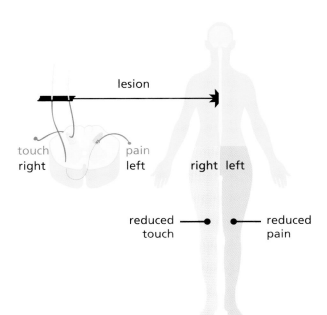

10.7 A **spinal hemisection** reduces touch sensation in the legs ipsilaterally and reduces pain sensation contralaterally, due to sectioning of axons from the anterolateral and dorsal pathways.

Let's engage in a bit of clinical diagnosis (figure 10.7). Imagine you have a patient who had a car accident and tells you that since the event, in her left leg she can feel touch but not any pain. But when she touches her right leg, she can only feel pain. By taking out your neurologist's pin (usually inserted inside the neurologist's reflex hammer) and carefully probing for sensations along her body, you map the affected dermatomes and conclude that she has an ipsilateral (same side) deficiency in touch and a contralateral deficiency in pain in her lower limbs. And, remembering your old professor's lectures on the dorsal and anterolateral pathway, you diagnose her with a lesion that cuts through the right half of her spinal cord (a right hemilesion) at a specific thoracic segment. And without having to do a CAT scan!

Central Projections

Feeling good about ourselves, let's follow the anterolateral pathway as it goes to the cortex—in fact, to several cortical regions (figure 10.8). This projection is anatomically complex. Depending on where axons

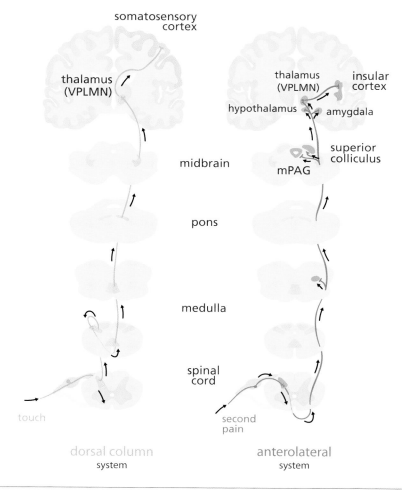

10.8 After decussating in the spinal cord, **anterolateral axons** project to the thalamus, hypothalamus, amygdala, and midbrain nuclei.

terminate, there are at least five major ascending anterolateral pathways: the spinothalamic, spinoreticular, spinomesencephalic, spinohypothalamic, and cervicothalamic tracts. All contribute to the central processing of nociceptive information and are thought to generate different types of pain sensations.

The spinothalamic tract is the most prominent one and is responsible for discriminative pain, whereby you can identify exactly what hurts. This is Buddha's first arrow. It includes the axons of nociceptive and thermo-sensitive neurons from the dorsal horn, terminating in the ventral thalamic nuclei, which then project to the primary somatosensory cortex to regions adjacent to the dorsal pathway.

What does this spinothalamic tract do? As good physiologists, to study the function of an organ or a pathway, we activate it or we lesion it, and this is how we can prove that it is causally related to a given function. If the function disappears when you lesion the organ, it was "necessary," and if the function appears when you stimulate it, it was "sufficient." Proving that something is necessary and sufficient is the hallmark of good causal science. Let's put this knowledge to good use: it turns out that electrical stimulation of the spinothalamic tract elicits the sensation of pain; conversely, lesioning this tract (via anterolateral cordotomy) results in a marked reduction in pain sensation on the side of the body contralateral to that of the lesion, as we discussed. So the spinothalamic pathway is causally related to pain processing.

What about the other four pathways? The cervicothalamic tract serves the same purpose for nociceptive information from the head. Meanwhile, the other tracts are thought to convey the affective and emotional character of the pain, as the second arrow. They terminate in the midbrain and hypothalamus and end up projecting to the amygdala and the medial thalamic nuclei; from there, they project to the cingulate and insular cortex, and all these structures are involved in processing emotional information and a lot of autonomic responses that are triggered by pain, as we will see in lecture 13. For example, during pain, our blood pressure increases, stress hormones are released, and our breathing changes.

So, in a nutshell, within the anterolateral system we have two main subpathways: a direct thalamocortical one, involved in sensory-discriminative pain information, and an indirect one, going through the midbrain, which is involved in affective-motivational pain. In fact, patients who have lesions in the cingulate or insular cortex still experience painful stimuli, but it doesn't seem to bother them. In other words, they know a stimulus is painful and where it's coming from, but they don't seem affected by it. It's the joined activation of all these areas that form our subjective perception of pain.

Referred Pain

I'm sure you've heard that a heart attack is often accompanied by pain in the left arm. Why is that? It turns out we also have pain receptors in our viscera, such as our heart or lungs. And these receptors send signals to the brain, but the somatosensory cortex somehow does not have a region devoted to them. So how does the brain process visceral pain information? It is sent out to the spinal cord and borrows the same circuitry that the spinal cord uses for the skin of the corresponding dermatome (figure 10.9). But that conservation of "real estate" comes with a problem: you can be fooled about where the pain is coming from.

Think about this: you have a patch of skin on your chest whose pain pathway goes to the dorsal root of the corresponding dermatome. But that secondary neuron in the pathway is also receiving the nociceptive pathway from the heart. So if this dorsal root neuron fires, it could be because either the left arm or the heart is injured—your brain doesn't know which. That's why you perceive the lesion in your heart as if it were a lesion in the skin, a condition known as *referred pain*. And a strong and abrupt pain in your left arm could be an indication of a potential heart attack.

I don't know why the brain would want to mislead you. I presume if you are a typical primate and don't have access to a surgeon, there is no point in knowing that one of your viscera is aching because there is really nothing you can do about it. But then why have pain fibers in your internal organs at all? I would love to hear if anyone has a good explanation.

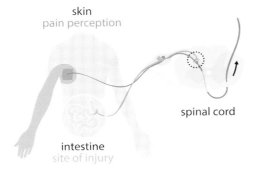

skin
pain perception

spinal cord

intestine
site of injury

10.9 **Referred pain** occurs because nociceptors from internal organs project to the same dorsal root neurons as those from the skin.

Phantom Pain

To make pain processing even more interesting, here's another mystery. Some amputees report pain in the missing limb! This condition is called *phantom pain* (figure 10.10). One potential explanation is that partly because of the amputation, the dorsal horn neurons are more excitable and "learned" the pain, leading to spontaneous pain. Until recently, limb amputation was performed under general anesthetic, assuming that this method would be sufficient to eliminate any memory of the traumatic procedure. Surgeons found, however, that even under general anesthesia, the spinal cord "experiences" the insult of the surgery. So, to reduce the risk of phantom limb pain, amputation now includes interventions that block dorsal horn neurons to ensure that they don't learn the pain and become excitable. General anesthesia is

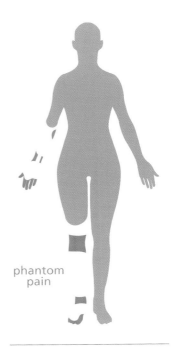

phantom
pain

10.10 **Phantom pain** occurs when pain arises from an amputated part of the body.

often supplemented with administration of an analgesic agent directly into the spine or a local anesthetic at the injury site.

Another explanation for phantom pain, not necessarily in conflict with the first one, is that it is generated by the cortex. Indeed, in the fMRI of phantom pain patients, you can see that parts of the cortex are active during the pain flare, as if they have become supersensitive. For example, areas of the cortex that were originally devoted to a missing hand or limb are now easily triggered by all kinds of stimuli, such as touching the lip. Interestingly, amputees *without* phantom pain do not show this abnormal cortical activity, which implies that the sensation of pain is generated by cortical activity, even in the absence of the part of the body that hurts. So it's all in your head.

Rather than a clinical curiosity, phantom pain, by demonstrating how pain can be centrally generated in the absence of any noxious stimulus, gives us deep insight into how pain is processed, as well as how perception works and how the cortex functions. Incidentally, it is often the case in medicine and neuroscience that the study of an exceptional clinical case or a particular odd behavior in rare species can reveal the logic behind the system. In these rare clinical cases of phantom pain, you can see directly how the perception of pain can be internally generated, even occurring without a stimulus (or a limb). That's why pain is "Kantian."

Descending Control

Now that we've climbed up to the somatosensory cortex and found out what it's up to, making our lives miserable with pain for limbs we don't even have, let's be positive and remember Dr. Livingstone. Why didn't he experience pain during his mauling? Well, it turns out that, in addition to afferent pathways that go up to the cortex, there are also efferent pathways that go back to the skin (figure 10.11). These pathways flow down from the somatosensory cortex to the amygdala and hypothalamus, then to the midbrain, and end up back in the dorsal horn of the spinal cord, where everything started.

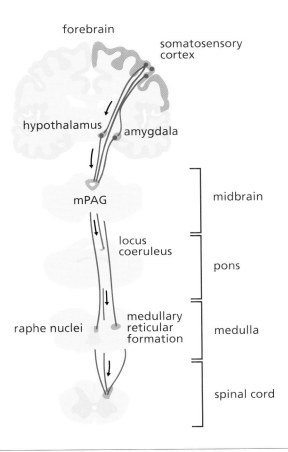

10.11 Incoming nociceptive information is modulated by **descending pathways** to the dorsal horn of the spinal cord.

As we discussed, there is a lot of interesting circuitry in the dorsal horn. Descending pathways to the spinal cord can regulate the transmission of nociceptive information, and they employ a wealth of different neurotransmitters that can exert both facilitatory and inhibitory effects on activity in the dorsal horn, effectively "gating" incoming pain information.

How does that work? Remember the C fibers coming from the skin, representing slow, burning pain? As you now know, they contact dorsal horn neurons projecting to the brain. But it turns out there are also other local neurons in the dorsal horn, called "interneurons" because they don't project outside the spinal cord, that are activated by the descending efferent projections. These interneurons release endogenous opiates that block the activity of the dorsal neurons receiving the C fibers, hence the analgesia (figure 10.12). Natural opiates from plants bind opiate receptors and thus highjack our internal system for pain control.

We have several endogenous opiates, including endorphin and enkephalin. Both opiates are peptides synthesized by neurons in the brain and are released as neurotransmitters. Once they bind to an opiate receptor, they open potassium channels, hyperpolarize neurons, or, in some cases, hyperpolarize synaptic terminals and also prevent transmitter release. That's why enkephalin release blocks the C fiber inputs in the dorsal horn, subduing or even eliminating all pain. This means that the brain can, in principle, switch off the transmission of nociceptive signals. That must be what happened to Livingstone. Why does the brain do this? One guess is that if you get mauled by a lion and want to survive, it's a good idea to forgo pain and screaming and crying for a while and focus instead

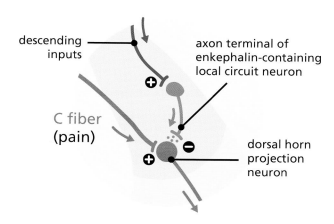

10.12 Descending inputs activate interneurons in the spinal cord that release **endogenous opiates** and block nociceptive inputs.

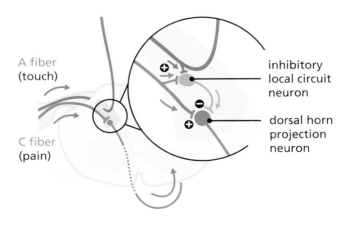

A fiber (touch)

C fiber (pain)

⊕

⊖

⊕

inhibitory local circuit neuron

dorsal horn projection neuron

10.13 Stimulation of touch afferents can activate local inhibitory neurons in the dorsal horn and **relieve** the effect of painful stimuli.

on running or fighting. That's exactly what happened to Livingstone, and he survived.

In addition to gating pain by descending afferents, local circuits within the dorsal horn can also modulate the transmission of pain to higher centers (figure 10.13). One example is the trick my grandmother used to use on us as children: rubbing the skin next to a lesion to make you feel better. The idea is that you activate low-threshold mechanoreceptors in the skin, which then synapse on inhibitory neurons in the dorsal horn, which contact the neurons receiving C fiber inputs and diminish the effect of nociceptor cells. Acupuncture probably works the same way, diminishing the sensation of pain by activating these interneurons.

▼ RECAP

The anterolateral system is a protective warning system that processes a lot of specialized somatosensory—and mostly noxious—information including sharp (first) pain, burning (second) pain, heat, cold, itchiness, and sensual touch, each of which has a dedicated pathway with specialized channels, receptors, and anatomical projections. This exquisite system of parallel pathways reaches the somatosensory and cingulate cortex, where nociceptive information generates the perception of discriminative and emotional pain. The cortex can generate pain internally in the absence of nociception, and can also control nociceptive information by gating its arrival at the spinal cord. Pain is therefore a good example of an internally constructed sensation, something that exists in our brains and can be triggered, or not, by the outside world.

Further Reading

Colapinto, J. 2009. "Brain Games: The Marco Polo of Neuroscience. A Profile of Vilayanur S. Ramachandran." *New Yorker*, May 4, 2009. A fascinating profile of one of the scientists behind phantom pain studies.

Julius, D. 2013. "TRP Channels and Pain." *Annual Review of Cell and Developmental Biology* 29, no. 1: 355–84. A review of the front end of nociception by a recent Nobel laurate.

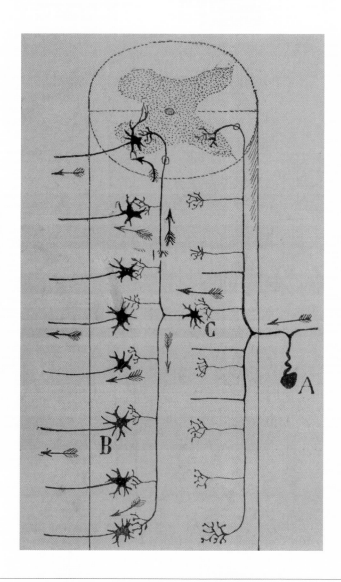

Current flow in unilateral reflexes. Neural circuits in the spinal cord generate most reflexes. Courtesy of the Cajal Institute, "Cajal Legacy," Spanish National Research Council (CSIC), Madrid, Spain.

LECTURE 11: REFLEXES

The mind . . . is the product of evolutionary processes that have occurred in the brain as actively moving creatures developed. . . . . Where does the story begin? What type of creature can we look to for support of this important connection between the early glimmerings of the nervous system and the actively moving, versus sessile, organisms? A good place to begin is with the primitive Ascidiacea, tunicates or "sea squirts." . . . The larval form is briefly free swimming and is equipped with a brain-like ganglion containing approximately 300 cells. This primitive nervous system receives sensory information about the surrounding environment. . . . These features allowed this tadpole-like creature to handle the vicissitudes of the ever-changing world within which it swims. Upon finding a suitable substrate, the larva proceeds to bury its head into the selected location and become sessile once again. Once reattached to a stationary object the larva absorbs—literally digests—most of his own brain, including its notochord. It also digests its tail and tail musculature, thereupon regressing to the rather primitive adult stage: sessile and lacking a true nervous system. . . . The lesson here is quite clear: the evolutionary development of the nervous system is an exclusive property of actively moving creatures.

—Rodolfo Llinás, *I of the Vortex: From Neurons to Self* (2002)

In this lecture, we'll learn

- ▶ How the knee-jerk reflex works
- ▶ How complex reflexes are assembled
- ▶ How neural circuits generate spontaneous activity

Motor System

So now that we have had a go at how sensory systems work, let's tackle the motor side of the brain for a few lectures. After that, we will try to put it all together into a coherent picture. Actually, to put it all together, and to come up with a general ideal of what the brain does, tackling the motor system is crucial because of the significant possibility that the original purpose of the nervous system is to make the animal move. And not just to move (even bacteria can move) but to generate and control movements in a way that is intelligent—in other words, to enable the animal to anticipate the future. As we introduced in lecture 1, the idea is that movement causes your environment to change, so you must be able to predict the future, and that can explain what the brain is trying to do: prediction–action. As the sea squirts in the opening quotation demonstrate, if you are not moving, you can spare your nervous system.

The way our brain generates most movements is from the top down: the *upper motor system*, led by the motor cortex, is involved in planning, initiating, and directing voluntary movements, which is helped by the basal ganglia and cerebellum. Then the motor commands are sent to the *lower motor system*—the brainstem and spinal cord—which controls the muscles but also can generate basic movements and reflexes independently of the cortex. This difference between upper and lower motor systems correlates with whether movements are voluntary or involuntary: cortically initiated movements are voluntary, whereas brainstem and spinal reflexes are not. Following this logic, the motor cortical areas are involved in decision-making and awareness and must generate some sort of idea of the self. Fascinating!

We'll start by discussing the lower motor system and then move upstairs to the upper motor system (figure 11.1). As noted, the lower motor system comprises the spinal cord and brainstem and is involved in the generation of reflexes and locomotion.

So what is a reflex? It's an involuntary response to a sensory stimulus. Essentially all muscles in the body are controlled by either the spinal cord or the brainstem (which you can think of as an extension of the spinal cord). The spinal cord receives sensory inputs, as we learned in the last two lectures, and also has motor neurons, so it has all the hardware to build these sensory-motor loops and generate reflexes locally. The spinal

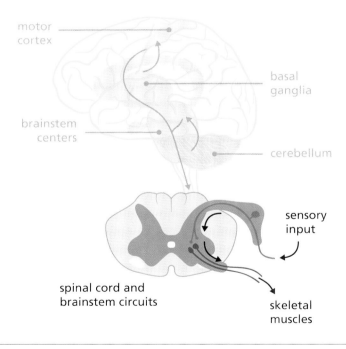

11.1 Most reflexes are built in the **lower motor system**, including the spinal cord and brainstem, which is controlled by the cortex, basal ganglia, and cerebellum.

cord also plays a role in locomotion—which is partly reflexive and partly controlled from above.

In this lecture, we will examine how the spinal reflexes use feedback circuits to implement basic control theory, which is one of the principles of neuroscience. Locomotion, conceptually very important for understanding the logic of the nervous system, is mostly built out of circuits called *central pattern generators*, which you know already. As I will explain at the end of this lecture, these CPGs could be considered the "soul" of the brain.

Muscles

To understand how the spinal cord generates movement, we need to first understand what muscles (or skeletal muscles, to be more precise) do. Muscles move the bones and different parts of our body by contracting or relaxing. That's all they do. Muscles are activated by the firing of motor neurons in the ventral horn of the spinal cord. They are divided into motor units, which are defined as a motor neuron and all the muscle fibers that this motor neuron innervates (figure 11.2). So, keep this in mind: a motor unit has many muscle fibers. Also, since the axons from each motor neuron branch in the muscle, the muscle fibers that belong to a motor unit are not next to each other. This positioning is likely meaningful, because this way the force exerted by the contraction of different motor units is spread throughout the entire muscle, generating a smoother movement (and also protecting the muscle and skeleton from being torn apart by strong local contractions).

It turns out that different motor neurons have different sizes and properties, and the muscle fibers they innervate also have different sizes and properties. So depending on the motor neuron you activate, the duration and strength of the contraction differs.

There are basically three types of muscle fiber: slow, fast fatigue-resistant, and fast-fatigable. The names say it all: slow fibers are slow to respond, they generate relatively little force and, importantly, do not get tired. Fast fatigue-resistant fibers are fast to act, generate more force, and

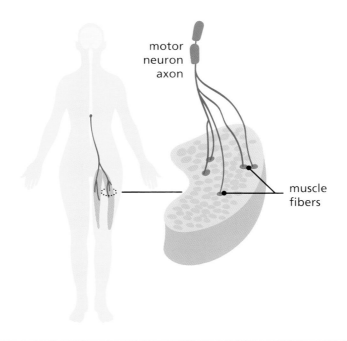

11.2 A **motor unit** is composed of all the muscle fibers innervated by a single motor neuron.

also partly resist fatigue. Meanwhile, the fast-fatigable fibers are fast to act and generate a lot of force, but they tire out quickly. Why do we need all these types of muscle fibers? Because our body recruits different proportions of fibers for different types of behaviors. For example, we use slow muscle fibers to stand, fast fatigue-resistant fibers to walk or run, and fast-fatigable fibers to jump (figure 11.3). That's why you can stand for quite a long time but get tired when you run.

Therefore, when exercising, we first engage the slower fibers, then the fast fatigue-resistant, leaving the muscle fibers that fatigue quickly in reserve until the end. This ordered recruitment is known as the *size principle*, because it correlates with the size of the motor neuron and the motor unit, and it makes perfect biological sense: only kick in the big guys that tire quickly when you really need them. That also means

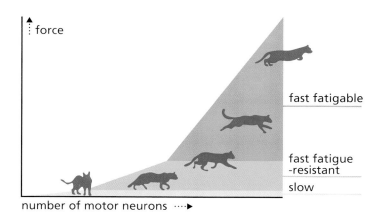

11.3 Motor neurons, and their motor units, are **recruited in order**, only engaging the stronger fast and fatigable fibers for the most strenuous behaviors.

the body has a whole repertoire of muscle fibers to be able to perform a certain task.

In fact, the human body evolved in the Paleolithic period, and is adapted to our past hunter-gatherer lifestyles. As animals, we are supposedly exceptional trekkers, meaning that we don't get tired when we go "walkabout," as Australian aboriginals still do. That's because we have the ability to "ride" on our slow or fatigue-resistant motor units for trekking, whereas other species use up all their fatigable fibers by running away from us. Because of this, we can outrun any animal, even animals that are much faster than us (like horses) and eventually follow their trail, track them, and exhaust them to death. For the argument of why we are essentially nomads in body (and also in spirit), read Bruce Chatwin's beautiful book *The Songlines*.

Spinal Cord

Let's go back to Sherrington, the father of neurophysiology (and my scientific great-great-grandfather, as I trained in his school). He argued that

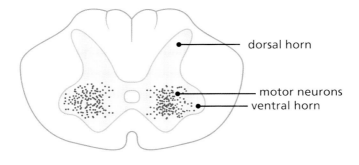

11.4 The **spinal cord** has a dorsal horn, with sensory neurons and a ventral horn with motor neurons.

motor neurons in the spinal cord are the "final common path," meaning that all behavior gets funneled through the activity of motor neurons. So whoever controls them controls all behavior. Motor neurons in the spinal cord—also named "*alpha*" motor neurons because they are the largest—, are located in the ventral horn (figure 11.4). Motor neurons controlling the muscles in our heads are located in the equivalent ventral position of the brainstem. The reason they are ventrally located has to do with development. As we discussed, the neural tube has a basic division of labor, with sensory cells located dorsally and motor cells ventrally. This structure is beautifully reflected in the positioning of the neurons in the adult spinal cord and brainstem: dorsal is sensory, ventral is motor.

Now the ventral horn motor neurons associated with a particular muscle are positioned in columns that run up and down the spinal cord. Unlike cortical columns, discussed previously, these columns of motor neurons are elongated clusters of neurons that innervate the same muscle. So in a way, we have a map of muscle innervation built in our spinal cord. Like in any good map, there is a basic topography: neurons innervating the musculature that is axial (closer to the center of the body) are located on the medial part of the ventral horn of the spinal cord. Conversely, the muscles that control the limbs (more lateral) are innervated by neurons that are located on the lateral ventral horn (figure 11.5). Again, there is

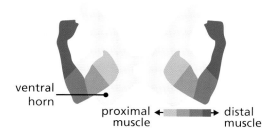

ventral
horn

proximal ← → distal
muscle · · · · · · · muscle

11.5 The **ventral spinal cord** has a topographic map of muscles, with motor neurons innervating distal muscles more lateral than neurons innervating proximal muscles.

a natural developmental explanation for this: the medial and laterally located nerves encounter the medial and lateral parts of the somites, respectively. Thus, in the spinal cord we again encounter maps, a basic principle of neuroscience.

I have to confess that there is actually a lot of local connectivity within the spinal cord, and we probably only know about 10 percent of what these interconnections are doing. While some motor neurons project to the ipsilateral side, both within a given spinal cord segment and up and down to the neighboring segments, other motor neurons decussate to innervate the contralateral side. What is the function of all these local connections? One thing we know they do is building motor reflexes. But to understand them, we have to explore something that happens within a part of the somatosensory system that we left for later, the one that innervates muscle: the proprioceptive system.

Proprioceptive System

The proprioceptive system monitors the position and tension of all of our muscles, tendons, and joints, and, by doing so, provides real-time information on the state of our skeleton and our body, something critical for generating any behavior. In fact, this monitoring is so crucial that axons that carry proprioceptive information (the group A class) are the thickest and most myelinated of all axons, which, as you now know by heart, makes them the fastest.

Proprioceptive sensory axons arise from structures called *muscle spindles*, which sit in the middle of muscle fibers (figure 11.6). These spindles have *intrafusal* muscle fibers, with proprioceptive group A axons that wrap themselves around the fibers and have stretch receptors in the membrane of their nerve endings. So when the muscle contracts, the intrafusal fibers also contract, and action potentials from these axons are sent to the spinal cord. Thus, group A axons monitor muscle tension continuously. (FYI, there are also similar terminals in tendons, in the so-called Golgi organs, which, rather than monitoring stretch and length, monitor tension and force.)

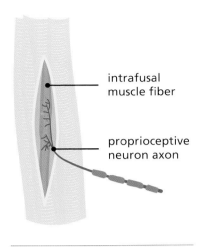

intrafusal muscle fiber

proprioceptive neuron axon

11.6 Muscle spindles are located in intrafusal fibers inside muscles and send sensory proprioceptive information to the spinal cord.

The bottom line is that your nervous system is constantly keeping careful track of the position, length, and forces of all of your muscles! Moreover, all this monitoring is done automatically, involuntarily, by your spinal cord, something that gives us an inkling of how much our nervous system is doing that we are not even aware of—or, for that matter, have voluntary control over.

Monosynaptic Reflexes

Now we are ready to dive into reflexes. Explaining how a reflex works was Sherrington's major contribution to neuroscience. Moreover, he argued that reflexes occur throughout the brain, and this reflex action model (or sensory-motor model), as we discussed in the first lecture became a recipe for how things work in the nervous system.

Let's first focus on simpler reflexes: the *monosynaptic reflexes*. They are so named because they involve a single synaptic connection in the spinal cord, a connection from a sensory neuron to a motor neuron. These simple reflexes are also called *stretch*, *deep tendon*, or *myotatic* reflexes.

A classic example is the knee-jerk reflex, which you may have experienced in a doctor's office or, if not, you can try it at home. Try this: if you are relaxed, sitting cross-legged, and someone hits (gently, please) the lower tendon of your patella (knee cap), your leg will jerk up in a small kick. It does that automatically, involuntarily, without your control. How does that work? When the tendon is hit, it pulls and stretches the leg muscle in the front of your thigh (the quadriceps), which activates and fires the proprioceptive axons in its muscle spindles (figure 11.7). These axons, bringing action potential trains, enter the dorsal horn of the spinal cord and go down to the ventral horn, where they contact the motor neurons that activate the quadriceps, the same exact muscle that was stretched. That is why your

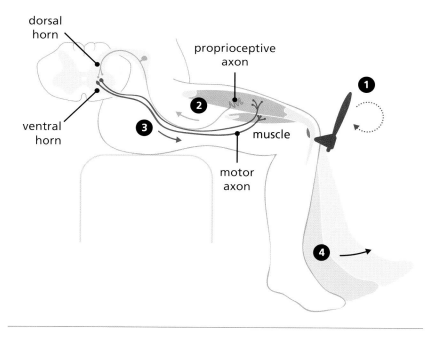

11.7 The **knee-jerk reflex** occurs when your patellar tendon is hit, stretching your thigh muscle. This activates proprioceptive axons that activate motor neurons, which contract the muscle and lift the leg.

muscle contracts and your leg pulls up. To make it more fun, the proprioceptive axons also contact inhibitory neurons that inhibit the antagonist leg muscle (the hamstrings, in the back of your thigh). So when you contract the quadriceps, you relax the hamstrings, like a well-oiled machine. If you think about all of this happening automatically, you'll be amazed.

But why do we have this weird reflex? What is the knee-jerk reflex good for, aside from helping you get admiring looks from family members as you demonstrate your newly acquired neurological skills? Stretch reflexes are some of the many built-in mechanisms to ensure that muscle function is smooth and coordinated and to maintain our upright posture. For example, the knee-jerk reflex probably prevents body imbalances created by changes in posture like when you walk. Its purpose is to keep our skeleton stable by compensating for a sudden stretch, maintaining the original length of the muscle. It's all about controlling your muscles.

In fact, the way the neural circuit underlying this reflex is set up is exactly a feedback loop (figure 11.8). You can rewrite the knee-jerk reflex

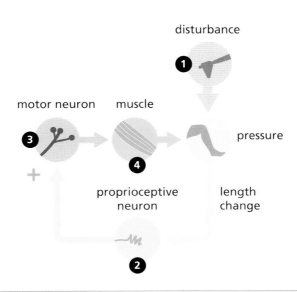

11.8 The knee-jerk reflex can be understood as a **control system** with a feedback signal that restores the original muscle length.

in the formalism of control theory, with a set point (the desired length of the muscle) and a feedback signal (the proprioceptive input). A new input generates a change in the system that compensates for it and preserves the system state. It's all about maintaining things the same, to achieve *homeostasis*, another term for "control."

Polysynaptic Reflexes

Now that we've got the knee-jerk reflex in the bag, let's tackle a more complicated reflex, one that involves not one but several synapses. We are still talking about that same leg of yours, but this time you are standing and happen to step on a pushpin. Ouch! From the last lesson, you know perfectly well how that pain sensation gets transmitted, but let's focus now on the motor side.

What happens when you step on a pin? You immediately pull back (flex) your leg and, at the same time, straighten (extend) the other leg so you don't fall. This action is called the *flexion-extension reflex* (figure 11.9). How does it work? Let's tease it apart. The afferent sensory neuron, carrying pain information from the foot, enters the spinal

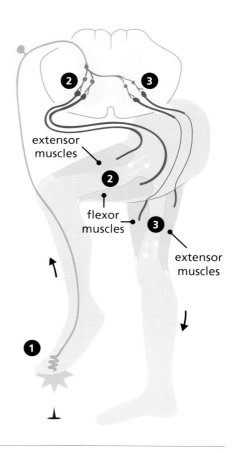

11.9 The **flexion-extension reflex** occurs when a pain afferent activates motor neurons to flex the ipsilateral leg, while extending the contralateral one.

cord via the dorsal root to synapse on spinal interneurons, which in turn project to motor neurons that activate the quadriceps (the same muscle responsible for the knee-jerk reflex), causing you to flex and lift your leg. At the same time, these interneurons make synapses on the contralateral side of the spinal cord, contacting the motor neurons that activate the hamstrings on the other leg, thus extending and straightening it.

These are all excitatory connections, but, as with the knee-jerk reflex, there are also inhibitory connections, in both the ipsilateral and contralateral spinal cord segment, that inhibit the antagonist muscles (the hamstrings of the leg you are lifting, and the quadriceps of the leg you are straightening). Voila! Isn't that beautiful?

No wonder Sherrington got carried away and extrapolated polysynaptic reflexes to more and more complicated behaviors. Like him, we can imagine that the entire nervous system, our entire behavior, is a table of reflexes, a gigantic machine in which a sensory input starts a chain of activity that flows through the system until it activates a pattern of motor neurons and generates behavior. From this point of view, the goal of neuroscience is to write down that entire phone book of reflexes, describing the receptive fields of sensory neurons, mapping their connections to the interneurons and the motor neurons, and the muscles. This is a very straightforward research program, and one that has been essentially at the core of neuroscience for the last century.

Locomotion

But there was a thorn—a pushpin, we should say—in Sherrington's model. What about behavior that occurs spontaneously, without sensory inputs? As you know, this question was asked by one of Sherrington's students, Thomas Graham Brown (that makes him my scientific great-grand-uncle). Let's revisit what Uncle Thomas was up to.

After he explored the knee-jerk and flexion-extension reflexes, the next logical step was for Brown to consider locomotion. If you analyze how a cat walks, you can see that it deploys a very precise motor pattern, much like a choreographed ballet, in which flexor and extensor are activated

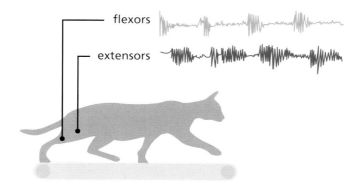

11.10 Locomotion involves a choreographed alternation of flexor and extensor muscles in the four limbs.

antagonistically in different limbs at different times (figure 11.10). This pattern of activation changes depending on whether the animal is walking, trotting, pacing, or galloping, engaging flexors and extensors in the four limbs in a coordinated way. So locomotion is no more than organized muscle activity in space and time.

Following Sherrington, Brown thought that the reason this happens is due to sensory feedback, since the movement of a limb would activate a next cycle of sensory stimulation from that limb, which would activate the motor neurons, triggering the next reflex, and so on. Thus, locomotion could be understood as a concatenation of reflexes.

Well, then comes Brown, who cuts all the sensory inputs into the cat's spinal cord by sectioning the dorsal root ganglion nerves. He also cut the connection from the brain to the spinal cord—leaving the spinal cord completely isolated in the body of the animal, with motor neurons coming out of it and no sensory neurons coming into it. And guess what happened? His cats were still able to walk even though they had no sensory inputs! To explain this, he proposed that the spinal cord itself can generate a pattern of firing that is spontaneous and cyclical and occurs regardless of the sensory input, and that generates locomotion. In fact, you can even extract the spinal cord from an animal, keep it in a dish, bathe it in the right media so it doesn't die, and you record from its motor neurons it still generates alternating patterns of activity, as if the animal were walking!

Central Pattern Generators

How does this patterned spontaneous activity occur in the absence of sensory stimulation? You may remember the half-center model from lecture 4, but let's revisit this again. Let's work out the circuit (figure 11.11). First, motor neurons that activate a given muscle on one side of the body connect with interneurons that inhibit the same muscle on the contralateral side. For example, activation of the flexor on one side will inactivate the flexor on the contralateral side. But, crucially, when inhibition ends, there is a rebound excitation that happens automatically. Many neurons are capable of this rebound response, which has to do with a property of sodium channels (their release from inactivation), which brings back the neuron to its action potential threshold. This rebound can also be aided if there are background excitatory inputs depolarizing the neuron. Importantly, these background inputs don't have to be sensory; they can just be unspecific inputs that are always on.

With this circuit in place, if you kick the system into action by, say, firing the flexor motor neuron in one side, the other is automatically inhibited. And then, when that inhibition ends, the contralateral flexor motoneuron

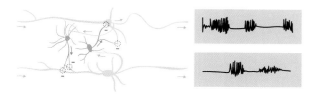

11.11 Using **reciprocal cross-inhibition**, ipsi- and contralateral motor neurons can generate alternating patterns of activity.

fires, which inhibits the ipsilateral flexor motor neuron, which then fires later, and so on. Just like a pendulum, a flip-flop. Brown called this his *half-center model*. This crossed reciprocal inhibition within the spinal cord, and the small delay caused by the inhibition until the rebound excitation kicks in, is what allows movement to alternate between sides.

This spinal cord network is an example of a *central pattern generator* (CPG); that is, a neural circuit that generates spontaneous rhythmic activity. It turns out that CPGs occur not only in the spinal cord of mammals but throughout the brain and all over evolution, all the way down to cnidarians. From the periodic contraction bursts of hydra, to the peristalsis in the stomatogastric ganglion of lobsters (remember figure 4.5?), to the swimming of leeches (figure 11.12), nervous systems are filled with CPGs.

OK, but who cares about hydras, lobsters, and leeches, what about humans? Our brains are probably full of CPGs. Do you remember Hans Berger's experiment using electroencephalography on his young son? Berger discovered that when his son's eyes were closed, there was strong oscillatory activity in his visual cortex, the alpha rhythm. This oscillation is a CPG. In fact, all oscillatory brain activity is generated by CPGs.

Why are CPGs so widespread? If you think about it, they generate spatiotemporal patterns of activity, which you can then put to good use to activate muscles and generate spatiotemporal patterns of movement. And that's all of behavior. But you can also use these endogenous patterns of

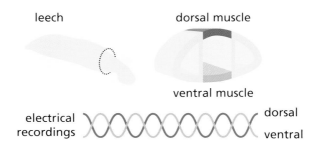

11.12 The **leech** spinal cord generates alternating pattern of contractions in the body muscles that help the animal swim.

activity for other things. As Llinás argues, during evolution, these CPGs (which do not necessarily need to be periodic or oscillating) can be repurposed for more abstract computations, whereby locomotor states now correspond to symbolic states in a model of the world such as memories, perceptions, ideas, and emotions. In fact, you can imagine that thinking is like mental walking, and see the spontaneous activity oscillations as the set of primitives to generate any desired spatiotemporal activity. This encephalization of CPGs could be responsible for the evolution of the mind. This has strong connections to neural networks, as the states of the CPG are mathematically homologous to the attractors of feedback neural networks.

▼ RECAP

The spinal cord receives somatosensory inputs from both our skin and from proprioceptive terminals in our muscles, joints, and tendons. It quickly processes them locally to generate patterns of reflexive activity in its motor neurons, which are relayed in a topographic manner to the muscle mass of the body. Reflexes are involuntary, outside our control, and are continuously engaged to enable a stable posture and smooth movements and serve a protective function to prevent injuries to our muscles and skeleton.

The study of the spinal cord circuits involved in simple spinal reflexes such as the knee-jerk reflex or in locomotion provides deep insights into the two alternative strategies that explain how the nervous system generates behavior. The first, exemplified by reflex arcs, which can be easily understood with control loops with a set point and feedback, argues that the brain essentially generates a long chain of synaptic steps triggered by a sensory stimulus, eventually activating motor neurons that activate muscles and generate behavior. The second viewpoint, exemplified by central pattern generators, which can be understood as network attractor states, argues that the brain is full of recurrent excitatory connections, building neuronal ensembles. These ensembles endogenously generate intrinsic spatiotemporal patterns of activity, which are then used to activate motor neurons, generating behavior, or activate other internal patterns, generating mental processes.

Both views are complementary, as each can explain a significant portion of our current neuroscience knowledge. Importantly, both views agree on the idea that the purpose of the nervous system is to generate movement, consistent with the evolutionary restriction of nervous systems to those metazoan species that can move. As animals move, it becomes essential for their survival to predict their future position in the physical world, and that likely provided the evolutionary pressure for the development of the nervous system and also for its fundamental role as a future-predicting machine.

Further Reading

Grillner, S. 2003. "The Motor Infrastructure: From Ion Channels to Neuronal Networks." *Nature Reviews Neuroscience* 4: 573–86. A comprehensive review of how the spinal motor system works.

Kiehn, O., and S. J. Butt. 2003. "Physiological, Anatomical and Genetic Identification of CPG Neurons in the Developing Mammalian Spinal Cord." *Progress in Neurobiology* 70: 347–61. A modern-day approach to dissecting CPGs in the spinal cord.

Llinás, R. 2002. *I of the Vortex: From Neurons to Self.* Cambridge, MA: Bradford Books, MIT Press. A lucid description of the importance of internal states for brain function. If you didn't read it when I recommended it after the first lecture, now you have no excuse.

Shepherd, G. M. 2003. *The Synaptic Organization of the Brain.* Oxford: Oxford University Press. A classic description of principles of brain circuitry. If you can find it, the 1990 edition is my favorite.

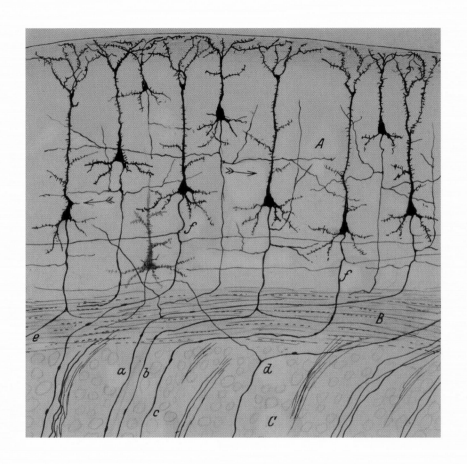

Pyramidal cells in the cerebral cortex. Pyramidal neurons in the motor cortex generate voluntary movements. Courtesy of the Cajal Institute, "Cajal Legacy," Spanish National Research Council (CSIC), Madrid, Spain. Pyramidal neurons in the motor cortex encode the idea of a movement.

LECTURE 12: MOTOR PLANNING

Further conceive, I beg, that a stone, while continuing in motion, should be capable of thinking and knowing, that it is endeavoring, as far as it can, to continue to move. Such a stone, being conscious merely of its own endeavor and not at all indifferent, would believe itself to be completely free, and would think that it continued in motion solely because of its own wish. This is that human freedom, which all boast that they possess, and which consists solely in the fact, that men are conscious of their own desire, but are ignorant of the causes whereby that desire has been determined.

—Baruch Spinoza, *Ethics* (1677)

OVERVIEW

In this lecture, we'll learn

- ▶ That the motor cortex has a map of body movements
- ▶ How the cortex encodes for the ideas and goals of movements
- ▶ How mirror neurons encode actions of other subjects

Upper Motor System

Are we free to act, or are all our actions predetermined by mysterious brain mechanisms? To scientifically answer Spinoza's question as to whether or not we have free will, we need to know more about the motor system. In the last lecture we discussed the lower motor system, including the spinal cord and brainstem. We learned that alpha motor neurons (the famous final common path) control essentially all musculature, and we worked out a couple of the spinal reflexes, which are conceptually important for the history of neuroscience regarding models of how our brains work. But reflexes are involuntary, and there is more to movement than that.

In this lecture, we will discuss the upper-level control of volitional movement and search for the source of our will (figure 12.1). We will meet the upper motor neurons, which the cortex uses to activate the spinal cord alpha motor neurons. You can think of the alpha motor neurons as the

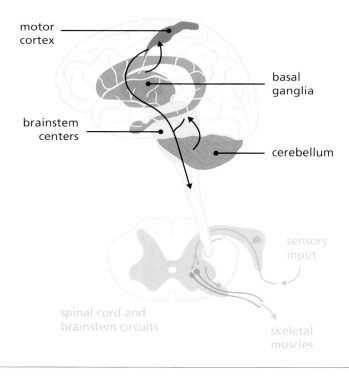

12.1 The **upper motor system**, composed of the cortex, basal ganglia, and cerebellum, generates all volitional movements.

keys of a piano that plays muscle units instead of notes. Now imagine that the motor cortex neurons are the performers playing the keys. In other words, the lower motor neurons are controlled by the upper motor neurons. Spinal motor neurons had an evolutionarily more ancient role in reflex arcs and then were "hijacked" later in evolution (particularly in primates) by the motor cortex.

It turns out that the main anatomical projection from the upper (cortical) to the lower (spinal) motor neurons is crossed. In previous lectures, we've talked about how all sensory information crosses over as it ascends to the cortex. There, some magic processing happens, decisions are made, and information is sent down via projections from the motor cortex to the spinal cord, which also cross over; they decussate. This contralateral pathway is known as the *corticospinal tract*, also called the *pyramidal pathway*. The nomenclature is confusing, because there are two pyramidal nametags in the brain, the one just mentioned and another one, our famous pyramidal neurons in the cortex, which you know well. We inherited this nomenclature from our past colleagues. To make things more confusing, pyramidal neurons from the motor cortex are the source of the pyramidal tract. But keep in mind that pyramidal neurons are essentially all the excitatory neurons in the cortex whether or not they are part of the pyramidal pathway. They are called "pyramidal" because the neurons look like a pyramid.

But why is the pyramidal pathway called pyramidal? If you look at the pyramidal tract axons coming from the motor cortex as they reach the medulla, you will see that in a transverse cross section, their tract looks like a pyramid. In fact, primates, including humans, have a particularly large pyramid-shaped tract, which reflects the major projection that stretches from the motor cortex straight to the spinal motor neurons that move the muscles in our digits. This pyramidal tract projection enables fine, skilled movements in our hands and fingers. You don't find such a direct projection in lower mammals, so I guess it is one of the abilities that make us human. Actually, that makes us primates, as these projections are present in primates. We are monkeys, skilled with our fingers.

But this control of motricity is not just exerted by the motor cortex talking to the spinal cord. Besides the pyramidal tract there are other tracts that go from the motor cortex to the reticular formation. The brainstem center is also involved, particularly in postural control, and both the cortex and the brainstem are helped by the basal ganglia and cerebellum, fascinating structures that deserve their own lectures.

Motor Cortex

So let's go into the motor cortex and take a closer look, being systematic. We are going to study the anatomy, the electrophysiological responses of the cortical neurons, and try to find out what they do by first lesioning them (to test if they are *necessary* for a given function) and then stimulating them (to test if they are *sufficient*). In other words, as good physiologists, we will explore the causality between these structures and their function.

The primary motor cortex is right in front of the central sulcus, which separates the frontal lobe from the parietal lobe (figure 12.2).

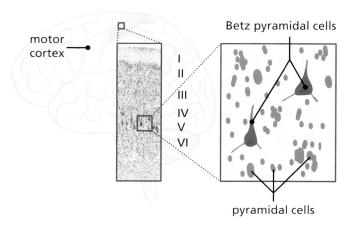

12.2 The **primary motor cortex** is located in the frontal lobe and has large pyramidal neurons.

It is interesting that the somatosensory cortex is posterior to the central sulcus and the motor cortex is right in front of it, on the other side of the street, so to speak, as if they were meant to talk to each other. Like in any other part of the cortex, if you view a cross section of the motor cortex, you will find layers, but also something special: the presence of very large pyramidal cells, the Betz cells. They are likely the largest neurons in the body, because their axons begin at the top of the cortex and travel through the spinal cord, where they contact the motor neurons that innervate your limbs. Axons from these neurons could be up to one meter long, which explains why the neurons are so large; they have to maintain a huge axon.

The motor cortex is divided into the *primary motor cortex* (M1), the origin of the corticospinal tract, and other motor cortical regions, which are known as *supplementary motor regions*, including also premotor cortex and prefrontal cortex. All of them are involved in motor planning.

Maps

What does the primary motor cortex do? Let's stick an electrode in there and find out. We can see that neurons fire action potentials when the animal produces a movement; for example, as a result of a wrist extension (figure 12.3). But, interestingly, if you look closely at the data, the firing actually happens just before the animal makes the movement. And once the movement begins, the motor neuron stops firing. Picture the upper motor neuron as the fingers playing the piano keys of the lower motor neurons. But because this cortical motor neuron is firing even before the movement, in this sense it may reflect the mind of the pianist, who wants to hit that key. In other experiments using electromyograms, we see that particular motor neurons fire with the activation of particular muscles. So there is a correlation between specific motor cortex neurons and specific muscles.

By systematically recording neurons across parts of the primary motor cortex, scientists found that M1 has a topography (figure 12.4). Different parts of the M1 are involved in movements of different muscles in

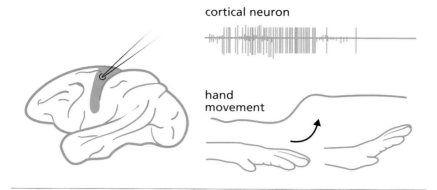

12.3 Neurons in primary motor cortex respond to selective **movements of body parts**.

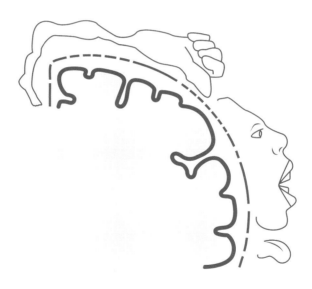

12.4 The primary motor cortex has an **ordered map of body movements**, with larger representations of parts whose movements are more important for behavior.

our body—it's not all scrambled but ordered. In patient studies, electrical stimulation of different parts of the motor cortex elicits small movements of different body parts. We have another homunculus in there! This motor map was first astutely deduced by the neurologist John Hughlings Jackson, who noted that in some epilepsies that progress through the body, the progression is stereotypical and reflects the physical progression of epilepsy across the motor cortex.

The motor homunculus is somewhat different from the somatosensory homunculus, right across the central sulcus, as the motor homunculus has an expanded representation of the parts of the body that generate the majority of useful motor movement. Medial in the motor cortex is the lower part of the body; then come the lower extremities then you have the trunk, and then you end up with a large representation of our upper extremities (hands in particular). Finally, more lateral in the hemisphere, we have a large representation of our face, ending up with our tongue.

Again, the principle is that the majority of the surface of the cortex is devoted to sensory or motor activities that are of particular interest to the animal. It is fascinating to examine this orderly motor representation of the body and realize how large is the size devoted to our hands (in particular, our thumb). In addition, a large part of the motor cortex is involved in our tongue (speech), and the rest of the body is sort of compressed. As with the somatosensory homunculus, the fact that our thumbs and hands have such a large representation in M1 goes back to the description of humans given by Heidegger, who defined us as toolmakers. We are skilled primates who use our hands to build tools, and that's written into our motor cortex. And we move our tongue a lot to talk.

But these motor maps are not as clear-cut as the maps we have seen in the sensory systems. For example, there is a mixture of neurons within a given territory that might respond to the activation of parts of the body outside of their supposed sections. Also, there are neurons that code for many things at once. So let's take these motor maps with a grain of salt and keep digging.

Population Codes

Besides motor maps, recordings in the primary motor cortex have also revealed a key principle of neuroscience: populations of neurons work together in groups, in ensembles.

In experiments with monkeys, researchers recorded the activity of neurons from the arm region of the motor cortex as the animal was performing a task. It had to respond to a light by moving the joystick toward it (figure 12.5). What the investigators found was very interesting: it turns out that many neurons in M1 exhibit directional selectivity, responding only in a particular direction of the arm movement, so they are "tuned" to a preferred movement direction. From these data a tuning curve can be produced, showing that tuning is broad, similar to what is found in the visual and auditory systems— cortex is cortex! In addition, as in the visual and auditory cortex, it turns out that, by systematically performing these recordings of the arm region, you find that different neurons have different directional selectivity. So you can record the activity of an entire population of neurons one by one.

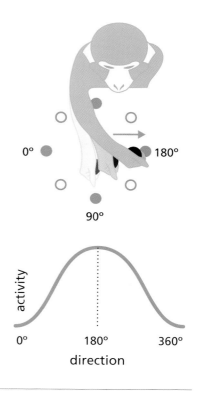

12.5 In this visuo-motor task, monkeys are trained to move a lever towards a light. Although each motor cortical neuron is not well tuned to the direction of the movement (bottom) the activity of the **neuronal population** exactly predicts the movement.

Now, if we then average the activity of all the neurons together, we can precisely predict the movement of the animal! In other words, by recording how each of the neurons fires and bunching them all together, we can predict the angle of the movement of the arm even before it moves. And that prediction is much more accurate than what the broad tuning of each individual neuron would allow. This analysis fits the hypothesis that individual movements are coded by a population of neurons, an ensemble, rather than just one neuron. The implication is that the job is done not by a single neuron but jointly by a group of neurons. Although each neuron is poorly tuned, working together as a population, they are very finely tuned since, by virtue of averaging, they can code the movement of the hand with exquisite precision. There are no special neurons; they all have the same say. It seems like a neuronal democracy in which every neuron has one vote.

Incidentally, the movements we are discussing are voluntary, so the outcome of this democratic vote going on inside your brain is your decision. This so-called population code fits right in with the idea of ensembles and attractors that I brought up in the earlier lectures.

Lesions

So what do these maps represent? If you record from these neurons in motor cortex, you'll discover that they don't really have a receptor field; rather, they have a *motor field*. If the receptive field can be described as the sensitivity of a neuron to a stimulus, you can think of the motor field as the tuning curve of a motor neuron, the movement that its firing corresponds to. Do these motor fields have any function? What is the function of neurons in the motor cortex? As good physiologists, let's figure this out by either lesioning or stimulating the neurons and observing their effect on the behavior of the animal.

Let's start by asking the question of whether the motor cortex neurons are necessary for movement. To test this, what happens when we lesion these neurons? For example, a monkey is trained to perform a task in which it has to pick up a morsel of food (figure 12.6). Now, if the pyramidal tract—the fibers that are coming from the primary motor cortex—is lesioned, the monkey can still pick up the food but is not as skilled at

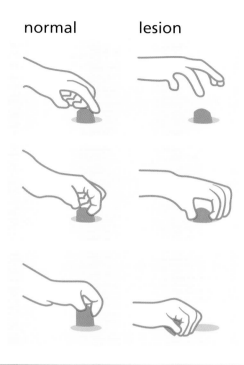

normal   lesion

12.6 After **lesioning the pyramidal tract**, motor behavior becomes unskilled.

performing the task. Instead of using its fingers, it pushes its whole hand to shove the food into it. This finding suggests that the primary motor cortical neurons are responsible for this fine, skilled behavior, because if you lesion them, you alter or damage the behavior. The monkey still moves its hand and the muscles are still activated, but the pattern of movement becomes disorganized. This poor performance resembles fairly accurately what happens in human infants. When you observe babies, you can see how their movements are quite imprecise, as if they lack a developed motor cortex and a pyramidal tract (which is actually the case).

Now that we have demonstrated that these neurons are necessary for the behavior, let's dig deeper into this causality. Are they sufficient for the behavior?

Stimulations

To test whether motor cortex is sufficient for movement, what happens if you stimulate its neurons? What you see is that you trigger peculiar nonrandom movements (figure 12.7). Based on this stimulation experiment, one could argue that the motor cortex is not only necessary but also sufficient for these complex movements, as it can trigger them. The causal case is closed.

But let's take a closer look at the movements that are generated. For example, if you stimulate a particular area of the motor map of a monkey, you can trigger movement of the arm to the mouth, as if the monkey is trying to put something in its mouth (an "eating-like" movement). But if you move to other parts of the motor cortex and stimulate those neurons repeatedly, you trigger movements from the arms that now go to the chest, as if the monkey is attempting to shield itself from injury. This looks like a protective or defensive movement. If you stimulate other

12.7 Stimulating motor cortex elicits **specific movements** that often have biological meaning (blue: protection; red: feeding).

parts of the motor cortex, you can even elicit more types of movement, such as climbing or reaching. Fascinating!

You may have noticed that these types of movements are very different from the mechanical "puppet string" type of movements that the spinal cord generates. These cortically triggered movements have biological meaning. Spinal motor neurons control the motor units in the muscle, and, because the muscle is always in the same place, every time it contracts, it performs exactly the same movement. Now, the motor cortex is something different; it codes for movement more abstractly. It's not as simple as moving the hand in exactly the same way. The hands can move in many different ways, but some of these movements have something in common—for example, it's the concept of moving something to your mouth. Cortically induced movements are more like ideas of movements—goals or intentions—rather than the actual specification of movement. This suggests that motor cortex is planning and generating the general concept of a movement and sending that information down for someone else to do the job. Motor cortex is abstracting, just like sensory cortex was.

Premotor Cortex

So far we've been discussing the primary motor cortex, on the central sulcus. Let's examine the area directly in front of the primary motor cortex: the premotor cortex. We find the same theme we learned for vision and audition: as we go up the motor hierarchy, we encounter more and more abstract properties (figure 12.8). Let's forget about the brain for a minute. One way to understand this structure and function is inspired by robotics (figure 12.9). If you were an engineer and had to build a robot that moves, you would first have to translate the intention to move into what's called *extrinsic kinematics* (the goal of the movement; for example, you have to catch a ball), then *intrinsic kinematics* (how you have to move the limbs to achieve the goal of catching a ball), and finally *muscle kinetics* (how you have to activate the different sets of agonist and antagonist muscles to fulfill the intrinsic kinematics). Until a decade ago, scientists thought that

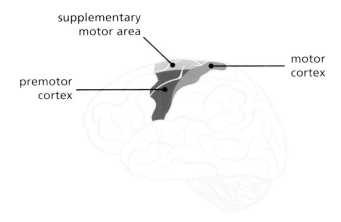

12.8 The **premotor** and supplementary motor areas extend into the frontal lobe, in front of the primary motor cortex.

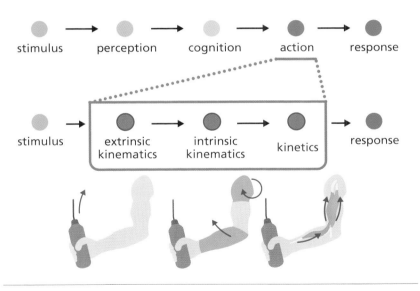

12.9 A **motor action** can be decomposed into sequential steps, where the intention of movement is translated into limb kinematics and muscle kinetics.

this is how the motor system works, in a very orderly fashion, with the extrinsic and intrinsic kinematics being computed in the premotor and motor areas and the kinetics specified in the spinal cord.

But in reality, if you record from different motor areas, or even other regions in the frontal and parietal lobes, you find that many neurons are involved in the entire process. Neurons that are closer to the actual muscle movement are located in the back near the parietal lobe, computing the kinetics (some do compute intrinsic kinematics); whereas the farther you move to the front of primary motor cortex, you find neurons computing extrinsic kinematics or intention. It's true that we have a primary motor cortical area with maps, but, particularly for the motor system, you don't find a clear dissociation of different tasks in different cortical areas—it's all mixed together.

This mixing of information gets even more interesting. The premotor cortex is strongly connected to parts of the parietal cortex that are involved in the "where" pathway (which you should remember from the vision lecture) and which receive both sensory and motor information. This is a fascinating area of study right now, trying to decipher the role of the premotor and parietal cortices and how they interact. But when you record from neurons in this *association cortex*, sometimes it becomes difficult to distinguish between sensory or motor responses or even joint sensory-motor. It seems that even the distinction between sensory and motor is starting to fall apart.

Although it's easier for us to study the brain by dividing it into different compartments—this part of the brain is doing the somatosensory job, that part is doing the motor job—in reality, even if different areas do have specific functional properties and as more research is done, we come to the conclusion that the entire cortex is working together as a unit. For instance, in our lab, where we focus on the primary visual cortex in mice, we are finding that many visual cortical neurons respond when the mice are running even if the animal is in the dark! So if you didn't know better, you'd assume that you are recording from a motor area rather than from primary visual cortex. That finding doesn't really fit the traditional model of the parcellation of the brain according to specific functions, as in a

Sherringtonian worldview, but it fits like a glove the idea that the brain could be a gigantic neural network in which every part of it is connected, in a couple of synapses, to every other part, tapping into that information, which is present throughout the network. We are ready for a new paradigm in neuroscience.

Free Will

Let's now stick our electrodes into the premotor cortex and try to figure out what it does. If you record from neurons there, it turns out that they also have visual and somatosensory receptive fields! That means they respond if you see things in a particular part of your environment or if you are touched on a particular part of your body. These visual and somatosensory fields are actually aligned, precisely matched. And, by the way, this also happens in the parietal lobes. In fact, the parietal lobe and the premotor cortex have maps of the world, but these are not the Cartesian maps of outside space that we use in science and in our culture. Instead, these are maps that are related to the body, and they are distorted.

Another thing that's interesting that you can see in recordings from the premotor cortex is neurons that are activated in a movement regardless of which of the limbs the animal is using. That's right: premotor neurons can fire when the animal moves an arm to the right, regardless of whether it's the left or right arm. It is fascinating to imagine that the premotor cortex has an even more abstract map of ideas of movement and is engaging the motor cortex to implement those ideas in a chain of command. Both contribute to building a sort of musical score of the activity of the muscles, which is downloaded to the spinal cord, which then moves the muscles via the alpha motor neuron in a particular order.

If you keep going up the chain of command, you get to the "general," which decides. You could have areas involved in the decision, or the free will, to carry out an action. In fact, there is an area, the supplementary motor cortex, which, if you lesion it, stops all movement. And not because the animal cannot move (everything else works fine in the motor system) but because it seems to lack the will to move. People who have lesions in their supplementary motor cortex often don't speak; they have mutism. They are perfectly capable of speaking, but they just don't seem to like to talk. Concepts like free will and the will of an action are directly connected to the fact that these acts are voluntary, and free will could be generated in these supplementary cortical motor areas.

Wait, did you say free will? Then, is there free will, or was Spinoza right? This is a fascinating ongoing debate. If you record the electrical activity on the frontal lobe of humans engaging in a decision-making task, you can detect an *event-related potential* (ERP) around 300 ms—the famous P300—which heralds the decision before the subject's reported conscious awareness that they want to make a movement. One interpretation of these results is that our free volition seems determined ahead of time by our brain. But wait, why shouldn't it be? Where else are we going to get the idea to move? From thin air? Bringing Spinoza into the twenty-first century, you could argue that what we call free will is a shorthand term (or illusion) that in reality encompasses all kinds of complex decision-making processes that are going on in the frontal and parietal cortices.

Mirror Neurons

Speaking of coding for the idea of a movement, listen to the story of this discovery, which by the way happened by chance, like the best of them. Researchers in Italy were recording from a particular neuron in the premotor cortical area of a monkey. The neuron fired when the monkey picked up raisins from the table. So far, so good. But then the researchers found that when they were still recording from that neuron, if another monkey happened to also pick up raisins, the neuron began to fire again. The neuron fired regardless of whether the subject monkey or the other monkey was doing the action. Moreover, the researchers found that the neuron also fired when the investigator himself picked up the raisin! That's how they discovered *mirror neurons*, so named because they mirrored the movement of the individual or an action performed by another individual (figure 12.10).

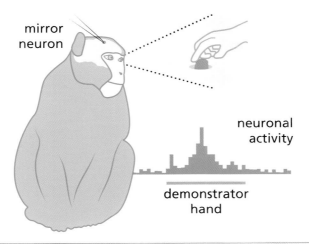

12.10 Mirror neurons fire when an animal performs an action or sees another animal perform the same action.

Following on other experiments in premotor cortex, the experimenters concluded that these neurons are encoding for intention to move. But that happens even if the movement is not performed by your body! Mirror neurons are encoding the abstract idea of the movement regardless of who does it; they are conceptual neurons. Mirror neurons have been found in humans and are present in both the premotor area as well as the parietal lobe and prefrontal cortex.

Because mirror neurons surpass the boundaries of the body, people have suggested they might code for other people's intentions and help to build empathy. What is empathy? It is when you feel in your own mind someone else's feelings. If you have mirror neurons for crying, when you see someone crying, maybe these neurons in your brain will fire as if you yourself were crying. Thus, if you have mirror neurons, you could have a model of other minds. These neurons could form representations of the minds of other monkeys or other people.

Because of this, mirror neurons could help build a theory of mind—the theory that there is an important social-cognitive skill that allows us to think about not only our own mental state but also the mental state of others. A theory of mind is at the heart of all of our social interactions. Some hypothesize that this system is defective in people with autism spectrum disorders. Autistic persons sometimes have problems with the concept of other minds, as if their mirror neurons were not working properly.

Cortical Remapping

Remember when we noted that the somatosensory cortex is very plastic? In fact, all of cortex is plastic, and in motor cortex we see this clearly again. The cortex is not like the spinal cord—it's always changing, and it is neuronal activity, reflecting our interaction with our environment, that is sculpting the circuits. Let me tell you about another experiment (figure 12.11), which resembles the experiments we covered in somatosensory cortex (figure 9.16).

First, researchers mapped the area of motor cortex that is devoted to each finger. Then they trained the monkey to perform a specific movement: use two fingers at the same time to access the morsel of food. Do you know what happened to the motor cortex map? It changed.

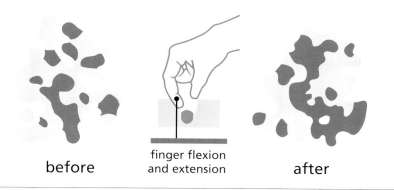

before

finger flexion
and extension

after

12.11 The cortical motor areas devoted to specific finger movement can grow after a repetitive motor task, demonstrating **cortical plasticity**.

The areas involved with those two fingers grew, while areas that responded to the rest of the fingers shrank. Here you have cortical plasticity in front of you.

As the monkey performs the task, its cortical area is changing. There is an increase in parts of the hand that are more useful to perform the task with a decrease in the areas that are less useful or when the behavior is no longer required. That means you can dial up and down the areas of the cortex depending on the behavior you are performing. If this happens in monkeys, it's happening in humans as well.

Brain-Computer Interfaces

I end this lecture with a technological advancement that will likely alter our lives soon: *brain-computer interfaces* (BCIs), also known as *brain-machine interfaces* (BMIs). These are neurotechnological devices that connect the brain to a computer. Imagine a paralyzed person who had a car accident, resulting in the complete severing of the spinal cord. Normally, to perform a behavior such as walking, motor information goes from the premotor cortex to the primary motor cortex to the spinal cord. But because the spinal cord is severed, the patient cannot move.

But with a BCI, you record activity from the patient's motor cortex and feed it directly to a computer (figure 12.12), which connects to a prosthetic limb, bypassing the spinal cord. After a training period, the patient can think about a movement, and that neuronal activity is decoded by an algorithm, which sends a command to move the prosthetic limb in the direction that the person wanted. This sounds like science fiction, but it works: a tetraplegic patient with a BCI and wearing a prosthetic exoskeleton delivered the initial kick in the 2018 soccer World Cup in Brazil.

It's important to note that BCIs work even though we have not yet deciphered the circuits in the brain and their *neural code*. The BCI electrodes are inserted into the motor cortex essentially at random, without knowing exactly which neurons are responsible for the movement. But even recording from the "wrong" neurons, after some training, the

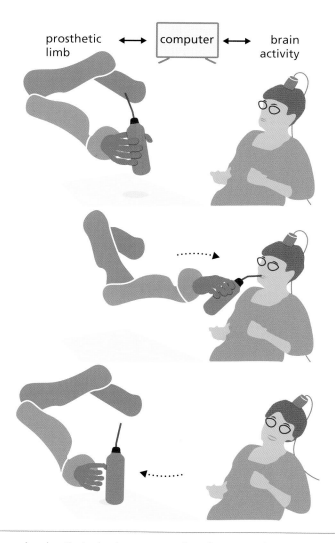

12.12 In paralyzed patients, **brain computer interfaces** record neuronal activity and connect it to a computer that can decode it and use it to move prosthetic limbs.

decoding algorithm works. We are not sure why it works, but it could be due to cortical plasticity. Through sensory feedback, the neurons that were probably encoding for something else are remapped to learn to code for the prosthetic limbs.

The Brazilian patient is an example of an invasive BCI, since you need a neurosurgeon to implant electrodes into the motor cortex. But many companies are developing noninvasive BCIs that can measure cortical activity with electrodes or optical sensors on top of the skull. Why is this important? In principle, you could record and decode cortical activity and connect it to a computer, providing a direct connection from our brain to the net. In addition, BCIs can not only record cortical activity but also modulate or stimulate it. This procedure is already being done with a variety of invasive deep-brain stimulation methods used in Parkinson's, stroke, and depression patients.

Now think of the implications: because of cortical plasticity, you could pipe information into a cortical area, and sooner or later, with training, the cortex would be able to decipher and use it. We know this works because, after a training phase, patients with cochlear and retinal implants are able to make sense of auditory and visual information.

So let's extrapolate to what will happen sometime in the future. Imagine wearing a noninvasive BCI that connects you with all the power of the internet and all its databases and algorithms, or with hardware devices that can serve as artificial sensory or motor systems. Sooner or later, this will happen. We will augment our sensory, motor, and cognitive abilities and transform ourselves as a species, incorporating technology as part of our body and our mind. This matter has serious ethical and societal implications, and we need to start thinking hard about how to integrate BCIs into our future world so it is done in an ethical way and enables us to preserve the essence of what it means to be human.

▼ RECAP

The motor cortex generates a motor plan and relays it to the spinal cord, which faithfully executes motor commands into muscle movements. This motor plan is assembled in steps, computing intentions (the goals, or why move) in prefrontal and parietal cortex, kinematics (the what, specific movement to generate) in premotor areas, and kinetics (the details of how to move the limbs or muscles) in primary motor cortex. These transformations are not just serial but also occur in parallel. In fact, perception, cognition, and action are all mixed in many cortical areas. Despite this mixing, one can find in motor cortex ordered maps of movements that are plastic, varying according to the use we make of them. This cortical plasticity enables the direct connection of neural activity to a computer with BCIs and the decoding of that information to control external devices. More generally, BCIs can be used as closed-loop systems to directly connect our brain to the net.

Further Reading

di Pellegrino G., L. Fadiga, L. Fogassi, V. Gallese, and G. Rizzolatti. 1992. "Understanding Motor Events: A Neurophysiological Study." *Experimental Brain Research* 91: 176–80. A review of how movements are generated from the discoverers of mirror neurons.

Georgopoulos, A. P., J. T. Lurito, M. Petrides, A. B. Schwartz, and J. T. Massey. 1989. "Mental Rotation of the Neuronal Population Vector." *Science* 243: 234–36. A pioneering paper desciding population coding.

Nicolelis, M. A., and M. A. Lebedev. 2009. "Principles of Neural Ensemble Physiology Underlying the Operation of Brain-Machine Interfaces." *Nature Reviews Neuroscience* 10: 530–40. A review of how BCIs work.

Yuste, R. et al. 2017. "Four Ethical Priorities for Neurotechnologies and AI." *Nature* 551: 159. A proposal to use human rights guidelines for the ethical development of neurotechnologies and BCIs.

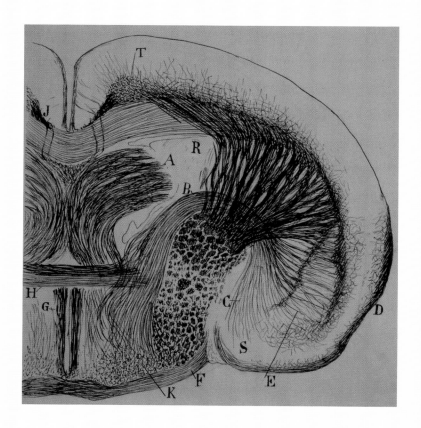

Frontal section of mouse brain. The massive cortico-striatal pathway, drawn as dark lines in the middle, sends motor information from the cortex to the basal ganglia, which select a particular motor action, while blocking all the other ones, thus ensuring that there is only one behavior expressed at any given time. Courtesy of the Cajal Institute, "Cajal Legacy," Spanish National Research Council (CSIC), Madrid, Spain.

LECTURE 13: MOTOR SELECTION

Dance is the hidden language of the soul.

—Martha Graham

OVERVIEW

In this lecture, we'll learn

- ▶ How basal ganglia choose one motor action at the expense of others
- ▶ How the brain generates reinforcement learning
- ▶ How alterations in basal ganglia generate movement disorders

Motor Selection and Control

Our movements reveal the inner workings of our brains, or our souls, as Martha Graham put it. But for any movement to occur, a critical part of behavior is the selection of which specific motor program to activate and which not to activate. While it may seem like a trivial task, it's not. Think about it: let's say you want to move your arm to scratch your nose. How do you ensure that you're not doing something else with the same arm at the same time, or using another part of your body whose movement interferes with scratching? Imagine you are trying to flee a predator and start scratching your nose. You would not last long in evolution.

It's a serious problem: behaviors need to engage exclusively one at a time. This is the type of issue that is so obvious that we take it for granted and never notice it, but it is true nonetheless. Moreover, somehow you need to keep tabs continually on which behavior works for the current plan and which doesn't—there is an extremely large number of possibilities for movement—and learn from it. All of this is done on the spot, constantly, right away while you are moving, to help craft behavior on the fly. Thus, you are unconsciously engaged in constantly monitoring and modifying motor patterns, which is needed in any control system that engages with the changing physical world.

How the brain performs this amazing feat is just beginning to be understood. Motor selection, motor learning, and motor control are carried out by two critical brain areas: the basal ganglia and the cerebellum. As discussed in lecture 11, they form two loops that modify the corticospinal pipeline (figure 13.1). You can think of them as serving to tweak

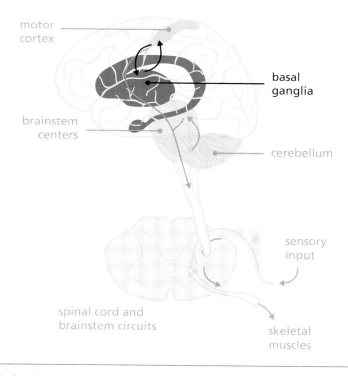

13.1 The **basal ganglia** are key parts of the motor system, and help select motor programs.

the cortical motor programs. But it's more than just a tweak: as we will see, there are serious consequences if they are injured.

Basal Ganglia

In this lecture I discuss basal ganglia, a collection of nuclei in the diencephalon that essentially form circuit loops from the cerebral cortex to the thalamus. Information from the cortex goes to the basal ganglia, then to the thalamus, and returns to the cortex. Thus, you can imagine the basal ganglia as part of a feedback loop, and the fact that this loop starts and ends in the cortex tells us that these ganglia must be involved in the modulation of cortical activity.

Ready for some anatomy? The basal ganglia are located in the cerebral hemisphere of the forebrain of the CNS (figure 13.2). As a reminder, the cerebral hemispheres include the cortex, basal ganglia, hippocampus, and

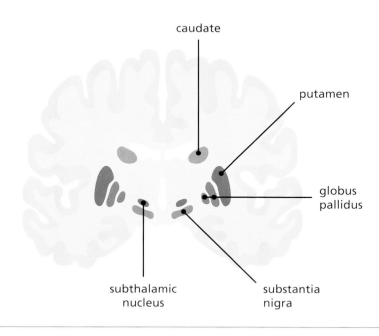

13.2 The basal ganglia are a **series of nuclei** located in the center of the forebrain.

amygdala. Basal ganglia look a little bit like a snake in the human brain; they appeared quite early in evolution, are buried under the cortical mantle, and extend from the front of the cerebral hemisphere and circle back.

The basal ganglia are divided into the *striatum* and the *pallidum*. The striatum is itself divided into the *caudate* and the *putamen*, and the pallidum into the *globus pallidus* and the *substantia nigra*. If you bear with me, the *substantia nigra* itself has two parts: the *pars reticulata* and the *pars compacta*. And, to top it off, there is a smaller part called the *subthalamic nucleus*.

From among all these nuclei, the main structures we will focus on are the striatum and pallidum. Why these names? Simple. If you look at them in a histological section, the striatum looks striated; in other words, striped. That is because axons are packed into large bundles (see Cajal's drawing). And the *globus pallidus* is so named because it looks pale in sections, and you can probably guess the color of the substantia nigra. So why does all this anatomy matter? It turns out that there are several loops, not just one, and when we lay them all out, we see that different parts of the basal ganglia are engaged with the cerebral cortex in different loops.

Feedback Loops

Let's take a look at these loops (figure 13.3). This will give us an inkling of what the basal ganglia are doing. First, the body movement loop—which, as the name says, is involved in movements in the body—starts in the motor cortex, in front of the central sulcus. It projects to the putamen of the striatum, which then projects to the lateral globus pallidus, which projects to the ventral lateral and ventral anterior thalamic nuclei, which then project back up to the central cortical region. Next, the oculomotor loop starts in an area within the frontal cortex called the *frontal eye field*. This area is a bit like the primary motor cortex but for eye movements.

We haven't talked much about eye movements in the book yet, but because humans are foveating primates, a significant amount of our motor computations have to do with moving our eyes. Remember also

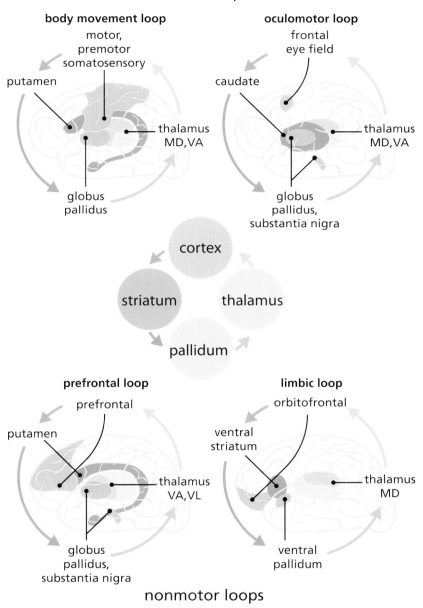

13.3 The basal ganglia form a series of **anatomical loops** that start and end in different parts of the cortex.

that our eyes are constantly tracking objects that we find interesting, so eye movements are involved in attention. So, from the frontal eye field, the signal goes to the striatum's caudate and from there to the pallidum, to a part of the globulus pallidus other than the body movement loop. From the pallidum, the loop continues to the thalamus and returns back to the cortex.

So far so good: the basal ganglia are involved in body and eye movements. But it turns out there are more loops, and they are not just for motor control. The first is the prefrontal loop, which is involved in all sorts of fascinating actions that we'll discuss later in the book, including calculating the future and evaluating the social environment. The prefrontal cortex is an area that is particularly developed in humans compared with other primates. That's probably what makes us human: we have this huge prefrontal cortex that gives us large foreheads compared with the skulls of other primates, whose skulls recede in the back. Well, this prefrontal cortex loop also goes first to the striatum, then the putamen/pallidum and the thalamus and finally returns to the cortex. So there is starting to be a method behind this madness, a reason behind the divisions of the basal ganglia. Each portion of the basal ganglia appears to be associated with a particular part of the cortex. The projections have a common design: cortex→ striatum→ pallidum→ thalamus→ back up to the cortex.

Now let's take a look at the final loop. This is called the *limbic loop* because it comes from the lower part of the frontal lobe, beneath and close to the midline. This portion of the cortex, sometimes called the *limbic cortex*, includes the amygdala, hippocampus, orbitofrontal, anterior cingulate, and temporal cortex. As we will see later in the book, these areas are connected to the brainstem and generate emotions.

In all four loops, we see the same anatomical theme: the homogeneous design of the basal ganglia loops. What this is telling us is that basal ganglia are controlling motor behavior, eye movements, the cortical part of our emotional brain, and whatever the hell is going on in the prefrontal cortex, you name it—intelligence, high cognition, imagination, and planning. Basal ganglia must be performing the same canonical computation and doing so with different cortical regions. But if you study figure 13.3

carefully, you'll also notice that the occipital, parietal, and temporal lobes are not involved in these loops. Mysterious, no? The basal ganglia seem to only deal with areas of the cortex anterior of the central sulcus.

So what is so peculiar about everything in front of the central sulcus? Well, everything in front of the central sulcus is either devoted to movement or associated with motor function, including emotions, which, as we will see, are also sort of motor. Another way to describe motor functions are those with which we interact with the world, whereas everything posterior is associated with sensory function, receptive, taking stock of the world. Makes sense, doesn't it? The brain may not be that complicated, after all.

Circuitry

Now let's take out our microscope and dive into the heart of the basal ganglia, first examining the striatum (figure 13.4). We will discover that the striatum is filled with so-called medium spiny neurons, which, as you recall, receive input from the cortex and project to the globus pallidus and the substantia nigra. And if you look closer, these neurons are chock-full of dendritic spines, hence their name. They receive inputs from cortical neurons, and each cortical axon probably connects to only one spine of a given medium spiny neuron. Then the axon continues and connects to another medium spiny neuron, and so on.

When Cajal first described the corticostriatal projection, he argued that this manner of connecting neurons was like telegraph wires strung on poles. You can imagine this as power lines connecting one pole to the next until it connects all the poles. Thus, in the ideal case, each medium spiny neuron would receive information from essentially every cortical neuron, and each cortical neuron would talk to essentially all spiny neurons. Why do this? It is the topologically perfect way to maximize the distribution of information.

We have seen this design principle in the olfactory cortex, and it will become even clearer in the cerebellum. It seems as if every striatal neuron is trying to receive information from as many cortical neurons

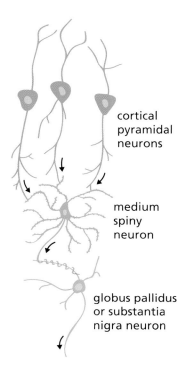

cortical
pyramidal
neurons

medium
spiny
neuron

globus pallidus
or substantia
nigra neuron

13.4 Striatal neurons receive inputs from many cortical neurons and project to few pallidal or nigral cells.

as possible, and, vice versa, cortical neurons are trying to distribute their output to as many striatal cells as possible. This, incidentally, also describes the connectivity matrix of a fully connected neural network, as we learned in lecture 5. So the likely reason these neurons are super spiny is because these dendritic spines are nature's way of achieving the maximum gathering of inputs. You can imagine that each dendritic cell receives input from a different cell and every spine represents one axon. So if the neuron had 300,000 spines, it could in principle collect information from 300,000 neurons.

But the opposite applies when we look at the connection between the striatum and the pallidum, since now each medium spiny neuron is connected to only one pallidal or nigral neuron. It wraps its axon around its partner's dendrites. And each pallidal or nigral neuron receives input from very few striatal neurons. It's a very interesting circuit. First, you are bringing in all these cortical axons and expanding the number of neurons that receive the information, yet in the pallidum, you're contracting the connectivity, focusing it and funneling all the information into individual cells.

Disinhibition

What are these connections doing? Here comes the surprise. The projections from the cortex to the striatum are all excitatory, but the projections from the striatum to the pallidum or substantia nigra are all inhibitory! Moreover, the projections from pallidum to thalamus are also inhibitory, and the final projection from thalamus to cortex is excitatory. If you add it all together in one picture, you'll see that this is an excitatory circuit built with two inhibitory steps (figure 13.5). How weird! Why? How does this work?

Let's take this one step at a time. Imagine first that when the cortex is at rest, the striatum is at rest. But if striatal neurons are not firing, then

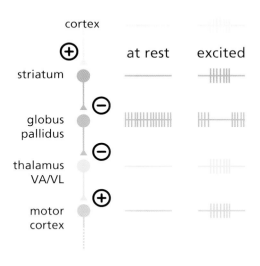

13.5 The basal ganglia loops have two **sequential inhibitory** connections, which make their overall effect excitatory.

the pallidus is not inhibited and will fire away, sending inhibitory signals to the thalamus. And because the thalamus is inhibited, the cortex won't receive excitatory signals, so nothing happens. But if we activate the cortex, it activates the striatum, which inhibits the pallidum, and then the thalamus is finally allowed to fire, as we turned off its inhibition produced by the pallidal neurons. At the end, if we activate the cortex, the cortex is activated back. This, then, is an excitatory loop that has two inhibitory intermediaries, which is called a *disinhibitory circuit*.

Two sequential inhibitory neurons in the circuit generate excitation. Mathematically speaking, two minus signs give you a plus. But why would you want to do this? Why would nature want to tap into what the cortex is doing, process it through these two negative steps, all to send the signal back up into the cortex? Didn't the cortex already have that information to start with? What a waste! Let's examine in more detail this inhibitory loop, which gets into the heart of the question: "What is the function of the basal ganglia?"

Before we answer that, let me point out that the pallidum, besides going to the thalamus, also projects to the superior colliculus. The superior colliculus is sort of similar to the thalamus but is particularly engaged in movements of the eyes and head. The reason we care about the colliculus is because the role of disinhibition in the basal ganglia is more clearly seen there. Imagine an experiment with a monkey (figure 13.6) in which we are recording from everything—the striatum (caudate), pallidum (substantia nigra), and superior colliculus—while the monkey moves its eyes. When the striatum fires, it inhibits the pallidum/substantia nigra, allowing the colliculus to fire. And when the colliculus fires, the monkey makes an eye movement! Somehow, the engagement of the basal ganglia loop has triggered that movement. This means you can trigger behavior if you stimulate this striatum.

But why not just stretch an axon from the cortex to the colliculus and save yourself all the trouble with the basal ganglia? The hypothesis is that the basal ganglia serve to keep the thalamus shut down all the time, as if by keeping it clamped that way, it does not engage the cortex unless something important happens. For behavior to be generated, somehow a few spikes need to escape from the cortex to the striatum, and after the

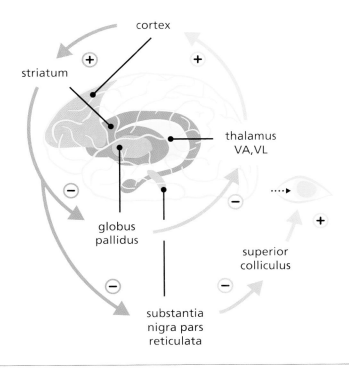

cortex

striatum

thalamus
VA,VL

globus
pallidus

superior
colliculus

substantia
nigra pars
reticulata

13.6 Activating basal ganglia disinhibits the colliculus and generates **eye movements**.

loop kicks in, a controlled movement can occur. That suggests it must be extremely important to not activate the wrong behavior. In other words, when you choose to perform a behavior, rather than simply turn it on, you have to release it from being shut off.

Moving is an exclusion problem. Behaviors occur one at a time, and if they didn't, chaos would result. As mentioned, you might move your arm to perform a task while at the same time trying to run from a predator. This is not good! You have to make sure that if you are typing on the computer, other movement ideas that are floating around in your brain are silenced. That sort of suggests that the brain is trigger-happy when it comes to generating behavior—at least the motor part of the brain. And the striatum is effectively shutting everything down except the lucky behavior that gets picked.

In other words, the function of this disinhibitory loop could be to release a particular behavior that has been chosen over all the other behaviors that are ready to jump in. So the way the basal ganglia regulate behavior is by choosing only one behavior and shutting down all the others. It's an example of a winner-take-all algorithm. By the way, that is still just a hypothesis, though it can explain the paradoxical double-inhibitory circuit.

Reward Prediction

Besides the behavioral exclusion I just discussed, another really interesting thing happens in the basal ganglia, and it could be related to how behaviors are chosen. Let me tell you the story of a discovery, another one that happened by chance. Wolfram Schultz was recording from dopaminergic neurons in the substantia nigra of a monkey, which was rewarded for performing the proper movement whenever its eyes moved toward a chosen direction. As you now know, neurons in the substantia nigra stopped firing when the monkey made an eye movement because, since they inhibit the colliculus, they released it from inhibition. But, interestingly, when the reward was given, the dopaminergic neurons started to fire shortly afterwards, as if they coded for the reward (figure 13.7).

Then things got even more interesting: if the reward was unexpected, the dopaminergic neurons fired a lot, but as the animal learned the task and the reward became predictable, the neurons stopped firing, as if they stopped caring. The better that the task was learned, the less the neuron fired to the reward. So as the animal became familiar with the task and knew a reward was imminent, the neuron displayed no altered firing.

The conclusion from this experiment is that dopaminergic neurons in the substantia nigra are not just coding for reward but for the *error of reward prediction*. What does this mean, "prediction error"? When the animal receives a reward but doesn't expect it, the neurons don't see it coming and respond to the novelty of the situation by firing. But once the reward is expected and the situation has become predictable and is under control, the neurons eventually do not respond.

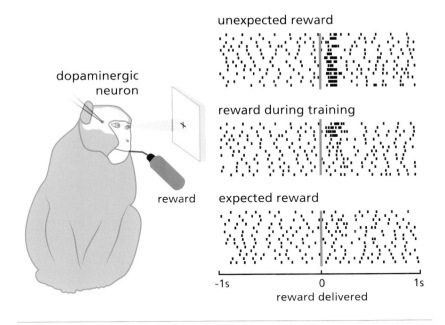

unexpected reward

reward during training

expected reward

-1s   0   1s
reward delivered

13.7 Dopamine neurons in the substantia nigra fire when the animal receives an unexpected reward, but stop firing if the reward is predicted, as if the neurons coded for **reward novelty**.

This finding suggests that these circuits are keeping track of what's familiar versus what's novel. They are coding for the novelty of the reward by releasing dopamine, which goes all over the brain and modifies neural circuits everywhere. Probably this is what happens when you get a "like" on Facebook: you get a little dose of dopamine if the "like" was unexpected, but if it's expected, no "reward" is generated through dopamine, and the neurons are not activated. A correct prediction is boring!

Monitoring error prediction is critical because you can use this signal to update the synaptic weights in a network. That is the basis of reinforcement learning, and it fits right in with the hypothesis discussed earlier of predictive coding. If the brain is trying to predict the future, it must measure the error between your guess and the reality and use that error signal to correct the model of the world. If that's the case, these dopaminergic neurons can be understood as coding for that error signal. This is learning by reward. The word "learning" and the basal ganglia are closely associated.

So now, thinking back about the role of basal ganglia: it's not just that you need to engage the basal ganglia loop to perform a behavior, but rather, you are opening this loop based on your prediction of what's going to happen. Thus, these are the two basic functions of the basal ganglia: to serve as a disinhibitory loop, as a way to release a behavior, and the calculation of the reward prediction error. By the way, reward prediction is actually a fantastic way to learn; some of the best computer algorithms learn to use this same error prediction model, copied from the basal ganglia.

Direct and Indirect Pathways

I'm sure you would agree that the basal ganglia have a complicated anatomy. Well, you are going to hate me for this, because it's going to get a little more complicated. But it's not too bad. You should now be familiar with the basic loop (figure 13.3). However, there is yet another loop (figure 13.8). The original loop is known as the *direct pathway* (cortex→ striatum→ pallidum→ thalamus→ cortex), but there is also an *indirect pathway* (cortex→ striatum→ pallidus→ *subthalamic nucleus*→ pallidus→ thalamus→ cortex). We stuck the subthalamic nucleus in the middle.

I mentioned the subthalamic nucleus at the beginning of the lecture. The projection from the pallidum to the subthalamic nucleus is inhibitory, while the projection out of the subthalamic nucleus is excitatory. Let's now do some accounting on this indirect pathway. The first connection is excitatory, positive (cortex to striatum); then we have one inhibitory, negative (striatum to pallidum); another negative (pallidum to subthalamic); a positive (subthalamic to pallidum); a third negative (pallidum to thalamus); and then a positive (thalamus to cortex). So that's three negatives, making it a negative, no? So the indirect pathway is inhibitory! Thus, this indirect pathway is the reverse of the direct pathway, which was overall excitatory. What's happening here?

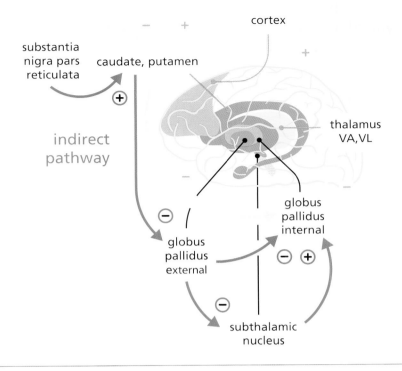

13.8 The **indirect pathway** in the basal ganglia counters the direct pathway, providing an inhibitory surround to it.

Remember the receptive fields in vision and the center-surround of the retinal ganglion cells? Now think of these two basal ganglia pathways as if the direct pathway were the center and the indirect pathway, the surround. The opposing actions of each pathway are similar to the purpose of the center/surround receptive fields for computing contrast. Remember, we defined contrast as the difference in luminance between the center and surround divided by the total luminance of the center plus surround. It's a differential contrast, to highlight the difference at the border of center and surround. So the purpose of this indirect pathway could be to create some type of *behavioral contrast* on the motor plan (figure 13.9).

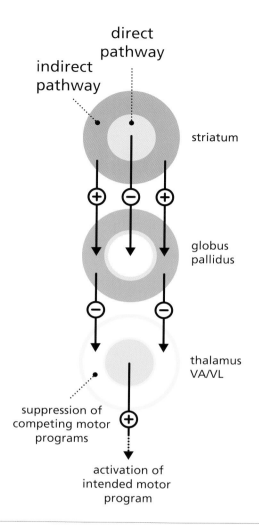

13.9 Through the opposite actions of the direct and indirect pathways, the basal ganglia enhance the difference in activity between chosen behaviors and unwanted ones, enhancing **behavioral contrast**.

You pick up one lucky movement (like the center) while suppressing all unwanted movements (like the surround). You can imagine each possible movement as flowing in a stream of computational space, as a particular algorithm that's propagating throughout the brain and is surrounded by this inhibitory surround to make sure the other behaviors don't happen.

For instance, let's say you want to play the piano and want to move only one finger, keeping the neighboring fingers inactive. To move one finger at a time, movement in all the other fingers needs to be canceled. If I want to move my middle finger to play the key (the center of this direct pathway), the indirect pathway (surround) would simultaneously prevent or clamp down all the other fingers. In this way, we could use the same computational trick that we discussed regarding visual contrast. Again, this is one of the basic principles of the sensory system.

We saw contrast computations in many sensory pathways, and it may also apply to the motor side. The brain could be computing contrast everywhere, including across the entire motor system. This could be a way to cleanly perform an action without interference. And that fits with the idea that the striatum is choosing one lucky behavior to be expressed from among the hundreds of behaviors that could be going through our minds at any given time. From all those choices, one gets picked in this winner-take-all scenario. This effectively shuts off neighboring behaviors in a computational sense—possibly maps of neighboring behaviors—through this indirect loop.

Diseases

Just as complicated machines often break down, the basal ganglia are affected in many diseases. First let's discuss Parkinson's. Another part of the substantia nigra comes into play here: the pars compacta, which has an excitatory role. I promise no more anatomy in this lecture! In a Parkinson's patient, the substantia nigra is essentially gone. It has degenerated and the loss of the dopamine pathways of the substantia nigra reduces the activity of the striatum, which leaves you with a diminished

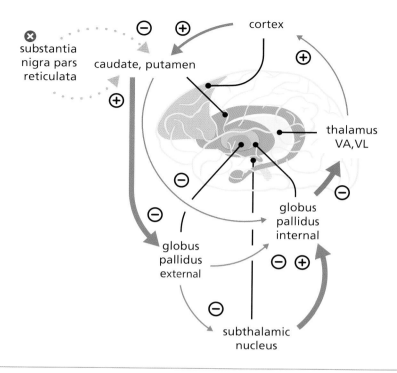

13.10 In **Parkinson's** disease, an impaired susbtantia nigra reduces striatal activity, which releases the pallidum, resulting in a reduced thalamic and cortical activity, with less movement.

input from the striatum to the pallidum (figure 13.10). And because of this diminished input, the pallidum is now less inhibited, which means that the thalamus is more inhibited, and there is less excitation to the cortex. The bottom line is less movement, and Parkinson's patients display diminished activity. They look rigid. They display tremor and other symptoms, but the major symptom is the lack of movement—lack of voluntary movement in particular. They seem to be frozen due to their hyperactive pallidum.

Patients with Huntington's disease display an opposite physiopathology (figure 13.11). A striatal degeneration results in a loss of medium spiny neurons in the striatum. So what happens without these neurons?

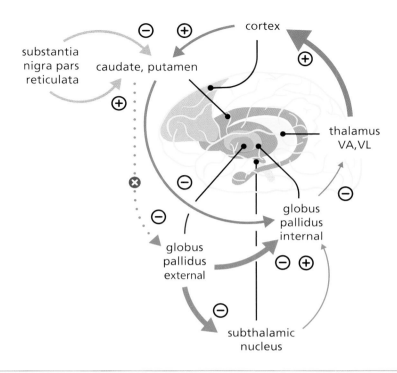

13.11 In **Huntington's** disease, an impaired striatum increases the activity of the subthalamic nucleus, which blocks the pallidum, resulting in enhanced thalamic and cortical activity, generating unwanted movements.

Following the indirect pathway, you have an increased activation of the subthalamic nucleus, which generates less inhibition. It's a bit complicated, but what matters is that through these alterations in the indirect pathway, in Huntington's disease, you essentially lose the inhibitory surround. That enhances motor tone and releases unwanted movements that were waiting in the wings. Indeed, many of the symptoms involve involuntary movements—sudden contractions of limbs (chorea), muscle problems, or head twitches.

Huntington's disease can be explained as motor programs that are normally suppressed becoming expressed, and this can happen also in other motor diseases. One example is Tourette's, which is also thought to arise from a disturbance in the cortico-striatal-thalamic-cortical circuit. I recommend that you watch *Motherless Brooklyn* (which is a fantastic movie in its own right) and pay attention to the symptoms that the main character displays. They could represent what may happen if you shut off the striatum, following the indirect pathway into the pallidum and subthalamic nucleus.

Huntington's and Parkinson's are on opposite sides of the spectrum, showing what happens when you change the delicate balance among the basal ganglia, thalamus, and cortex.

▼ RECAP

With the disclaimer that basal ganglia are complicated, and we are still in the early days of working out their function, we do know from animal experiments and patients that basal ganglia are critically important for movement. One hypothesis is that they act as a movement selector, involved in the focused selection of behavior.

In addition, basal ganglia are also involved in reinforcement learning. And the anatomical loops of the basal ganglia suggests that they are involved in some form of control. To build feedback control, you need to predict the error and make corrections accordingly to minimize that error. This fits well with the idea that the basal ganglia are one of the brain areas involved in the brain's reinforcement learning in terms of motor movements. That's a critical job if you want to be good at making a model of the world, using it to predict the future, and acting accordingly.

Finally, not only do alterations of basal ganglia have major motor impacts in one way or another, but they can also have nonmotor consequences, as you would expect from the different types of anatomical loops. I didn't go into this, but patients with Parkinson's disease also display nonmotor symptoms. Indeed, memory, cognitive, and emotional symptoms are also present in degenerative diseases, which should come as no surprise, given the basal ganglia's special relation with the prefrontal cortex. In fact, the role of the basal ganglia in nonmotor loops could be even more important than the motor ones.

Further Reading

Burguière E., P. Monteiro, L. Mallet, G. Feng, and A. M. Graybiel. 2015. "Striatal Circuits, Habits, and Implications for Obsessive-Compulsive Disorder." *Current Opinion in Neurobiology* 30: 59–65. https://www.sciencedirect.com/science/article/abs/pii/S0959438814001706. An interesting discussion of how basal ganglia are involved in habits.

Graybiel, A. M., and S. T. Grafton. 2015. "The Striatum: Where Skills and Habits Meet." *Cold Spring Harbor Perspectives in Biology* 7, no. 8: a021691. https://cshperspectives.cshlp.org/content/7/8/a021691. A comprehensive review of the heart of the basal ganglia.

Schultz, W. 2015. "Neuronal Reward and Decision Signals: From Theories to Data." *Physiological Review* 95, no. 3: 853–951. https://journals.physiology.org/doi/full/10.1152/physrev.00023.201. Review of the discovery and implications of dopamine reward signals.

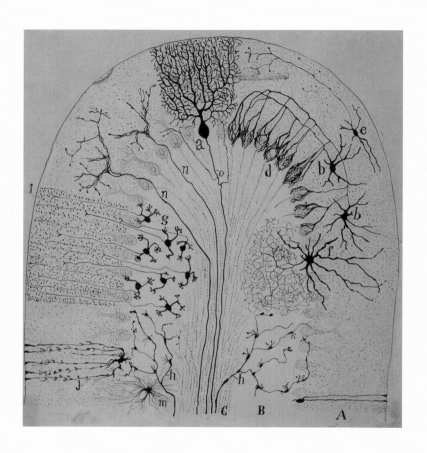

Transversal section of cerebellum of a mammal. Like a crystal, the neuronal microcircuits in the cerebellum are a beautiful example of a modular structure that repeats, and whose computations generate precise learned behaviors. Courtesy of the Cajal Institute, "Cajal Legacy," Spanish National Research Council (CSIC), Madrid, Spain.

LECTURE 14: MOTOR CONTROL

Criticism may not be agreeable, but it is necessary. It fulfils the same function as pain in the human body; it calls attention to the development of an unhealthy state of things. If it is heeded in time, danger may be averted; if it is suppressed, a fatal distemper may develop.

—Winston Churchill (1939)

OVERVIEW

In this lecture, we'll learn

▶ How the cerebellum is built with a crystalline neural circuit
▶ How the cerebellum computes motor errors to fine-tune behaviors
▶ How alterations in the cerebellum lead to deficits in motor learning

Cerebellar Loop

Let's reflect on our last lecture. It's not that we fully understand what the basal ganglia do. But at least we have hypotheses that make us think about motor systems in a different light. And notice how, by looking at patients, we have gathered some of our best insights into what basal ganglia do. So now, what about the cerebellum (Latin for "little brain"), this major structure of the hindbrain which is also part of the motor system?

Let's apply the same approach as we did for the basal ganglia. What happens if you lesion the cerebellum? Well, it turns out that you can completely remove the cerebellum of a mouse and there are only relatively small differences in behavior! If a person has a lesion in the cerebellum, you also only see some small but interesting effects. For instance, say you have a tumor in your cerebellum or had a car accident that caused the loss of part of it through whiplash-induced mechanical trauma. You'll end up with a variety of symptoms, including instability in your stance and gait, loss of precise and finely skilled movements, and learning new movements also becomes very difficult. Interesting.

Let's take a closer look at this peculiar part of the brain. What is the structure of the cerebellum? How does it connect to the rest of the brain? What are its potential functions? How does its structure implement those functions?

Inputs

As we have done in previous lectures, to understand what the cerebellum does, let's first learn about its structure, its anatomy, because we know more about that than about its function. Let's start by remembering the position that the cerebellum and basal ganglia have in the motor system (figure 14.1), as if they were modulating, controlling, or instructing the descending pyramidal pathway. And, if you take a closer look at the connectivity of the cerebellum, you see something similar to what we just saw in the basal ganglia, with a loop pathway that goes from the cerebral cortex through intermediate steps to the cerebellum, to the thalamus, and back up to the cortex (figure 14.2). Another loop! Now this one is a little different from that of the basal ganglia, as it also sends outputs to the brainstem and spinal cord.

Now let's zoom into the cerebellum. The cerebellum is part of the hindbrain near the pons and medulla oblongata, so in a way it is a relatively ancient brain region. In fact, fishes have a proportionately larger cerebellum compared to us. We do have a pretty sizable cerebellum, but it's not nearly as large as our cerebral cortex; on the other hand, the cerebellum

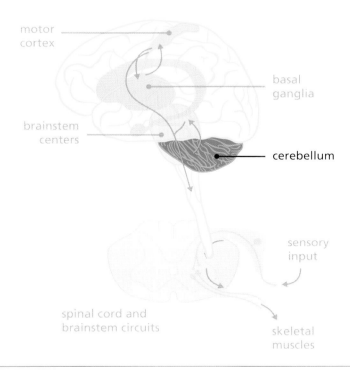

motor cortex

basal ganglia

brainstem centers

cerebellum

sensory input

spinal cord and brainstem circuits

skeletal muscles

14.1 The **cerebellum** is strongly connected to the motor system and controls skilled behaviors.

has more neurons that the cortex. It is packed with a particular type of cell, called *granule cells*. Lots of them—about fifty billion! So many, in fact, that there are more neurons in the cerebellum than the rest of the brain put together!

Now let's cut the cerebellum sagittally (parallel to the midline) to examine its different parts. Its inner core is formed by several nuclei connected to the brainstem (figure 14.3). And the surface of the cerebellum is similar to the cerebral cortex, in appearance and name. To make it a tongue twister, you learned about the cerebral cortex, and now we have the *cerebellar* cortex to deal with. Cortex means "bark" in Latin, so it refers to the surface of the structure. The cerebellar cortex, like the cerebral cortex, is highly folded into lobules called *folia*. These folia resemble

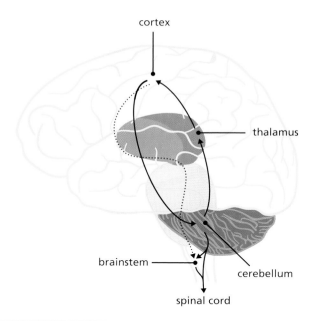

14.2 The cerebellum receives **inputs** from the motor cortex and sends **outputs** to the thalamus, brainstem and spinal cord.

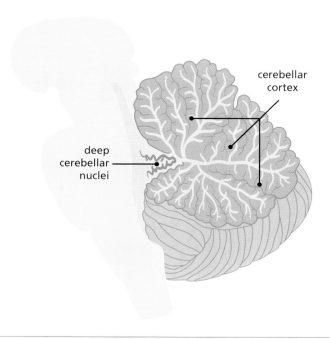

14.3 The cerebellum has **internal nuclei** and an external highly folded **cerebellar cortex**, maximizing surface area.

a cauliflower, with deep invaginations that are even deeper than those found in the cerebral cortex. It's as if nature is cramming as much cerebellar cortex as it can fit inside our skull. So, you can see the same folding principle as in the cerebral cortex: increase the surface area and shove it all inside the skull.

Cerebellar Cortex

Now let's take the cerebellar cortex out, unfold it, and extend it on a surface. You will see that it has three parts (figure 14.4): the *vestibulocerebellum* at the bottom, the *spinocerebellum* in the midline, and the third, the *cerebrocerebellum*, to the sides. Why do they have these names? Very simple: they describe three different loops, just like the basal ganglia have several different loops.

The vestibulocerebellum is connected with the vestibular nuclei and controls our balance. We didn't discuss the vestibular sense in earlier lectures: it is the part of the inner ear where your sense of position

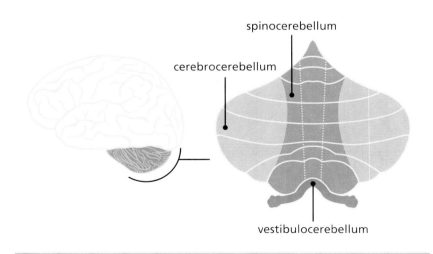

14.4 The **cerebellar cortex** has three areas, connected with different parts of the motor system.

with respect to gravity is computed. This information is all fed through this loop. So if you have lesions in the vestibulocerebellum, you have equilibrium problems. You'll tend to tumble. Incidentally, when you drink too much alcohol, your cerebellum malfunctions, perhaps related to the fact that most of the neurons of the brain are up there, so it could be the most sensitive part of the brain to alcohol. And this is why roadside sobriety tests involve having the driver walk in a straight line. The police officers are, unbeknownst to them, testing the state of your vestibulocerebellum!

The spinocerebellum has to do with the spinal cord, as the name suggests. And you can probably figure out what is connected to the cerebrocerebellum, no? In fact, the parts of the cortex that project to the cerebellum include a large portion of the frontal cortex and also parts of the parietal cortex (figure 14.5). This also tells you that the cerebellum is not only going to be involved in motor function (in the frontal cortex) but also in computing some spatial representation of the world (in the parietal lobe), which speaks to the cerebellum's role in motor learning. So, the cerebellum has loops not only from the cortex but also the spinal cord and brainstem that regulate movements of the body.

Maps

Let's keep zooming into the cerebellum and dive into the cerebellar cortex. It turns out that there is a little map in there, another homunculus! This one is a motor map, where specific parts of the cerebellar cortex are activated when the person is performing particular movements (figure 14.6). Although it looks like a homunculus, the way the primary sensory and motor cortices did, it has different distortions of body proportions. In this case, the mouth is not huge, as we saw in the somatosensory homunculus.

In the cerebellum, the lower limbs occupy more space, with the head taking over a relatively small area. The hands and both upper limbs are represented, but there's not a clear representation of individual digits. This representation suggests that the cerebellum has a more evolutionarily

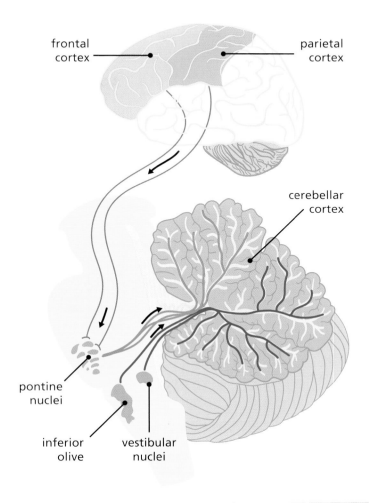

frontal
cortex

parietal
cortex

cerebellar
cortex

pontine
nuclei

inferior
olive

vestibular
nuclei

14.5 The **cerebrocerebellum** receives inputs from the frontal and parietal cortex via the pons and also inputs from the olive and vestibular nuclei.

ancient homunculus—one that is more similar to a lower vertebrate than a primate. What about the other parts of the cerebellar cortex? Are there also maps? There seems to be also a topography, but it is not so clear. In fact, I bet you there are all kinds of maps, but we just haven't figured them out yet.

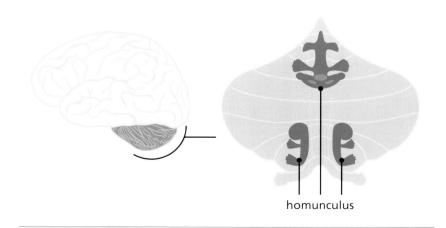

homunculus

14.6 The cerebellar cortex has mapped **representations** of the body.

Outputs

By now we know how the cerebellum is organized anatomically. The cerebellar cortex receives information about our balance and position with respect to gravity (from the vestibular nuclei), eye movements (from the brainstem), skeletal muscle (from the spinal cord), and voluntary movement (from the cerebral cortex). Looks like it is tapping into all the information you need related to moving the body and changing its posture. But which parts of the brain does the cerebellum talk to?

The cerebellum is receiving the information from these mentioned areas, processing that information and relaying it back to the cortex, brainstem, and spinal cord (figure 14.7). Specifically, the cerebellum has an upward projection to primary motor cortex and premotor cortex via the deep cerebellar nuclei and the ventral lateral complex thalamus. And it has a downward projection into the brainstem and spinal cord, through the deep cerebellar nuclei. So the cerebellum must be doing some sort of transformation to motor information and sending it back precisely to the parts of the nervous system that are executing motor actions. What are these transformations?

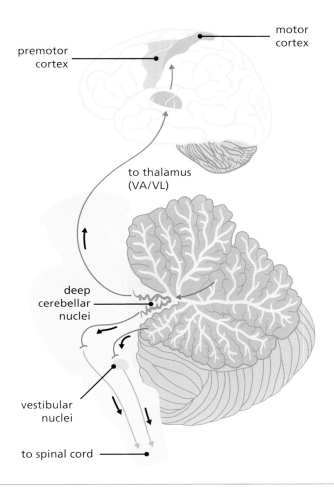

14.7 The cerebellum sends **outputs** to the thalamus and motor cortex and also to the brainstem and spinal cord.

Microcircuit

Let's continue to zoom in, and now let's look at the anatomy on an even deeper level. And—get ready to encounter one of the most fascinating corners of the brain! Although the function of the cerebellum is still not completely understood, a lot is known about its microcircuits. In fact, more than a hundred years ago, Cajal essentially figured it out and drew circuit diagrams of the cerebellar cortex (figure 14.8). Our heroes here are the Purkinje cells: presiding over the cerebellar cortex is a layer of these enormous, flat neurons that look like trees and receive hundreds of thousands of afferent connections from different neurons. Yes, I said hundreds of thousands. These are probably the most connected neurons

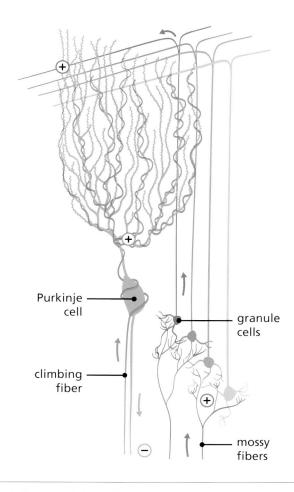

14.8 The cerebellar cortex is formed by repetitions of a basic **modular circuit**.

in the brain, which may explain why they are the largest neurons. Purkinje cells, which happen to be inhibitory, collect inputs and then generate the sole output of the cerebellar cortex. Their axon forms inhibitory synapses onto the cerebellar nuclei, which then send excitatory axons outside the cerebellum.

The vast majority of inputs to Purkinje cells come from granule cells: tiny cells located under the Purkinje cell layer. These granule cells, which are excitatory (and, as you now know, make up more than half of the neurons in the entire brain), have axons called *parallel fibers*. Each parallel fiber axon bifurcates, forming a T, and each of the branches of the T makes contact with tens of thousands of Purkinje cells! Since the Purkinje cells have flat dendrites, they intersect the bundle of parallel fibers, as if they are trying to catch as many of them as possible. As you can see (figure 14.8), this configuration resembles electrical wires running through the countryside, where the Purkinje cells are the towers and the parallel fibers are the power lines.

This resemblance is not a coincidence—this geometry achieves two things: it minimizes the number of contacts between the wires and supporting poles while maximizing the number of contacts with different poles (Purkinje cells, in this case). In fact, not only are Purkinje cells flat and stacked, but parallel fibers cross them at exactly 90 degrees. This is as good as it gets if you are a granule cell and want to contact as many Purkinje cells as possible while ensuring that you contact each Purkinje cell only once. Beautiful! Bravo! Nature at its best! But why would you want to do that?

Let's continue to pull the thread: who connects to granule cells? They receive excitatory axons, called *mossy fibers*, which originate in the pontine nuclei, the spinal cord, and the brainstem reticular formation. Each mossy fiber synapses on hundreds of granule cells (and on cerebellar nuclei neurons). Purkinje cells also receive excitatory climbing fibers, which originate in the inferior olive, a midbrain motor nucleus. Climbing fibers wrap themselves around the dendrites of Purkinje cells, forming hundreds of synapses, and stimulate the Purkinje cells so strongly that every time a climbing cell fires, a Purkinje cell fires.

But just as each Purkinje cell receives inputs from up to hundreds of thousands of parallel fibers, it receives input from just a single climbing fiber. Climbing fibers and parallel cells illustrate two opposite connectivity principles. In the case of the climbing fibers, we see a single fiber going to a single Purkinje cell to form many synapses, so that the information is funneled into only one cell. Parallel fibers, on the other hand, represent a single fiber going to many neurons to make a single synapse, so that the information is spread out to an entire population. These are two basic strategies for engineering circuits and neural networks, as we have already learned.

Finally, in addition to Purkinje cells and granule cells, the cerebellar cortex has basket cells, which are inhibitory neurons that surround the Purkinje cells. Basket cell axons look like little baskets, which beautifully encapsulate the cell body of Purkinje cells.

Motor Learning

The cerebellar microcircuit that we just described is like a crystal (figure 14.8). This pattern of Purkinje cells, mossy fibers, and granule cells with parallel fibers and basket cells is repeated throughout the cerebellar cortex and in all species. Nature is building modules of circuitry that are repeated over and over again, as if it is printing them. But why? What does this cerebellar microcircuit module do? There are lots of theories, but it is still not completely clear.

Let's think about it, trying to simplify the details. The cerebellar circuit is a loop with two input branches: inputs into the cerebellum activate the Purkinje cells in tandem, through the climbing and the parallel fibers, which are both excitatory. Purkinje cells do some "magic" and send an inhibitory output, which gets relayed on to the motor part of the brain. What is that magic? It seems to have to do with learning. Purkinje cells are masters at synaptic plasticity. They compare the two inputs they receive, and if those inputs arrive at the same time, they weaken (depress) that particular parallel fiber synapse. This makes the Purkinje cell fire less the next time around, when it receives the exact same input.

But because the Purkinje cell is inhibitory, firing less will activate more of whatever part of the brain is receiving its input, reinforcing that motor action or program. In other words, it is another disinhibitory loop, like with the striatum. This smells like feedback control, as if the cerebellum is modifying and tweaking the motor program based on some learning algorithm. One possibility, if you remember control theory, is that Purkinje cells are comparing movement prediction with actual measurement of the movement, computing an error—a motor error—and sending this information back. If this is indeed the case, these cerebellar circuits that are beautifully laid out in front of us would be used to tweak future movements for motor learning.

But what exactly is motor learning? Let's take a look at one funny experiment (figure 14.9). A neuroscientist goes to a bar. She talks with other fellow neuroscientists, and they get into the typical argument about whether the brain is settled in its ways or can still learn after the critical period. They make a bet: can you still play darts if you wear a set of glasses that distort your vision? Is your visual system set in stone or is it plastic? She says it is plastic, and they say it isn't. Let's see what happens.

Like a good scientist, she first gathers baseline data on her own behavior by throwing darts at the dartboard and accurately hitting the target. Then she puts on glasses that shift her vision to the right and finds that her darts are now systematically off target, falling to the side of where she is aiming. No learning has occurred. But as she continues to throw darts while wearing the glasses, her throws become more accurate, and eventually all the darts land on the target again. It looks like she has learned to compensate for the shift in vision! Then she removes the glasses, and her throws are off target again but now in the opposite direction, as if she is still compensating for the glasses. After a few more throws without glasses, her aim is back to normal and she wins the bet, as she has demonstrated that visual learning in adults continues to occur; in other words, the brain is still plastic.

This example illustrates the learning, or adaptation, of the connection between vision and motor programs. This adjustment happened automatically; our neuroscientist didn't have to calculate how far to the left to

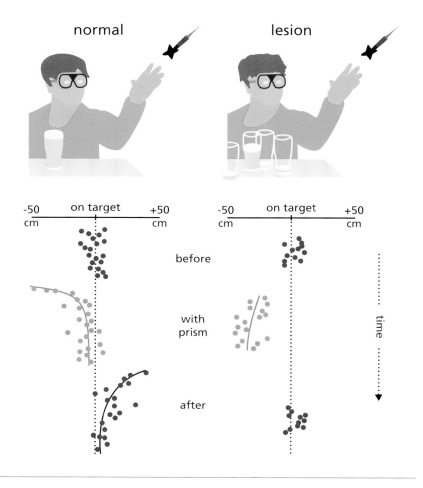

14.9 Patients with cerebellar lesions have deficits in **motor learning**.

throw the dart. Somehow her motor program is being tweaked on the fly to achieve the desired outcome.

But now look at what happens if our hero, with her hard-earned winnings, buys everyone beers and, after one too many, decides to perform the goggles trick again. But this time, her cerebellum is not functioning due to her high blood alcohol level. Guess what happens? No adaptation! The cerebellum is necessary for the adaptation: patients with lesions in the cerebellar cortex cannot adapt (figure 14.9, right). They see perfectly well, they can throw the darts on target, but if they put on the glasses, they never improve.

Lesions

So we have learned that the cerebellum is involved in motor learning, and lesions studies confirm this. Let's continue to examine the effects of cerebellar lesions. In fact, let's be systematic and classify cerebellar patients depending on where exactly they have their lesion. If we do this, the symptoms make sense.

Let's first take the vestibulocerebellum. Lesions in it cause loss of equilibrium—alterations in gait and nystagmus (unstable eye movements). If you have a lesion in the spinocerebellum, you end up with loss in muscle tone, muscle weakness, and dysmetria, over- or under-reaching when you have a voluntary movement like grabbing a glass, particularly after having too much to drink (figure 14.10). Meanwhile, lesions in the cerebrocerebellum generate ataxia (i.e., incoordination of movement) and also impairment in volitional and highly skilled movements, like playing the piano.

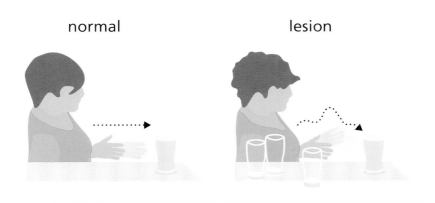

normal   lesion

14.10 Cerebellar lesions are associated with deficits in **skilled movements**.

Nonmotor Functions

All of these lesions make perfect sense and would appear to close the case: the cerebellum is involved in the coordination of movements and motor learning, perhaps by computing a motor error so that postural responses can be continuously adjusted. But there is also intriguing evidence that it may be involved in other types of learning. It turns out that the cerebellum lights up in fMRI scans of patients during memory tasks! This unexpected finding suggests that the cerebellum is involved in memory, and this could be due to the ancient role that the motor system could play in cognitive functions. This may be related to another fascinating idea: that the cerebellum, by generating temporal oscillations, is the clock (or metronome) of the brain, helping to synchronize movements of different parts of the body or cognitive processing.

Indeed, if you were an engineer building a moving robot, you would need to have a clock that synchronizes everything. For example, the muscles in your leg need to be activated along with the muscles in your arms to coordinate all the postural reflexes. One way to synchronize people is by creating a rhythm (think about dancing). It turns out that all responses of the cerebellum are oscillatory! The cerebellum oscillates at frequencies of 8 Hz, which is the same frequency as the normal small tremor and occurs throughout the body. In this view, alterations to the cerebellum could lead not only to motor deficits but also deficits in the synchronization of nonmotor information. In fact, it has been suggested that the cerebellum is affected in neurocognitive syndromes and also in autism.

▼ RECAP

As Churchill said, criticism is absolutely necessary to improve things, and motor control, or motor "criticism," is as important as motor planning and generation. To be able to operate in the real world, which changes all the time, an animal has to continuously update its posture and adjust its movements. It's not merely a prediction game but an action game as well. The cerebellum, an ancient structure with a crystalline microcircuit, appears to be making this continuous adjustment by taking motor predictions about our balance and position with respect to gravity (from the vestibular nuclei), eye movements (from the brainstem), skeletal muscle (from the spinal cord) and voluntary movement (from the cerebral cortex); comparing them with what happens after we act; computing some sort of error signal; and sending it back to the motor system in a loop. It is tapping into all the information you need related to moving the body and changing its posture.

Thus, while the cortex is the site of perception and behavior planning, the cerebellum may be responsible for motor prediction and the correction of motor errors through learning. Cerebellar lesions, though not fatal, clearly delineate the specific role of the cerebellum in motor control—and most likely also in neurocognitive processing and memory.

Further Reading

De Zeeuw, C. I., and C. H. Yeo. 2005. "Time and Tide in Cerebellar Memory Formation." *Current Opinion in Neurobiology* 15, no. 6: 667–74. https://www.sciencedirect.com/science/article/abs/pii/S0959438805001571. A review of the role of cerebellum in motor learning and memory.

Llinás, R., ed. 2012. *The Cerebellum Revisited*. New York: Springer-Verlag. A classic book from a pioneer in the study of cerebellar function, arguing for its role in timing and oscillations.

Wolpert, D. M., R. C. Miall, and M. Kawato. 1998. "Internal Models in the Cerebellum." *Trends in Cognitive Sciences* 2, no. 9: 338–47. A discussion of the idea that the cerebellum builds internal models of the motor system and uses them to compute prediction error.

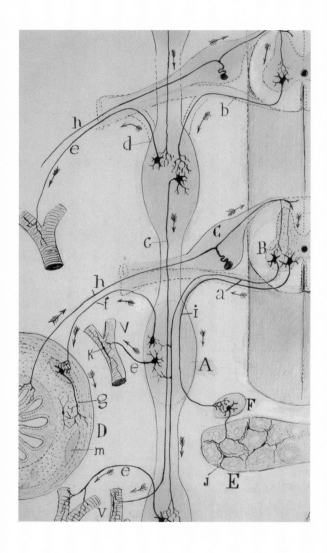

Sensory and motor pathways of the sympathetic system. The autonomic system implements a complex set of specific responses that are engaged during emotions. Courtesy of the Cajal Institute, "Cajal Legacy," Spanish National Research Council (CSIC), Madrid, Spain.

LECTURE 15: EMOTIONS

The best and most beautiful things in the world cannot be seen or even touched. They must be felt with the heart.

—Helen Keller, Correspondence (1891)

OVERVIEW

In this lecture, we'll learn

- ▶ How emotions mix cognitive, homeostatic, and motor responses
- ▶ How the autonomic system generates "fight and fright" or "rest and digest"
- ▶ How the enteric nervous system operates semi-independently of the brain
- ▶ How the hypothalamus regulates hormonal secretion

At this point in the book, we sort of understand how the brain could work, how it's put together during development, how sensory systems measure the world like good engineers, and how the motor system uses that information to plan smooth behavior and adjust it based on its mistakes, using control theory. Now let's take a turn and tackle a more complex issue: emotions, which are the things that really matter to us, that move us, and that we live and die for, as Helen Keller put it.

Actually, I shouldn't say that emotions are complex, but less well understood. I'll bet you that, at the end, they are simple and easy to understand. The reason that they are not well understood is because many of what we call emotions are not clearly defined scientifically to start with. I should warn you that we are now on unstable ground scientifically. Measuring emotions is not like measuring photons in the retina. What exactly is love? How do you measure it? What exactly is an emotion or a feeling?

The general agreement is that "emotions" are physiological responses of the body to a particular stressor—either positive or negative—and that "feelings" are the conscious cognitive processes that accompany emotions. But how does the brain control or trigger emotions or feelings? Talking loosely, one could argue that emotions are unconscious and are generated subcortically, whereas feelings are conscious and involve the cortex. However, physiological responses are often similar for emotions and feelings, so a lot of this is really all mixed together.

Even if we don't quite understand what emotions and feelings are, as we will see in this lecture, there is a significant and specific anatomical hardware that is involved in emotional response. So let's look at the anatomy and then review the function, as we always do. First, we will detour to discuss two parts of the nervous system that we left behind: the autonomic and the enteric nervous system. Then we will zero in on the hypothalamus and the limbic system, and then we will try to put it all together.

The autonomic and enteric systems, which, together with the dorsal root ganglia, form the peripheral nervous system (box 2.1), are technically also part of the motor system, as they can generate motricity and behavior. But they are also fundamentally involved in internal states, so they have a direct link into our cognition. The autonomic and enteric systems are probably evolutionary remnants of an ancient nervous system. and they actually resemble the nervous system of cnidarians and invertebrates with a ganglionic structure and also heavy use of neuromodulators and peptides. Yet the autonomic and enteric nervous systems live inside our body sometimes acting independently from the rest of our nervous system as if they were a different animal, cohabiting our body with us. Fascinating!

Autonomic System

Let's start with the autonomic nervous system. Why do we say that it is motor? Because of the way it is built: neurons from the brainstem or the spinal cord (*preganglionic neurons*) activate autonomic neurons in ganglia that are located throughout the body, and these *postganglionic neurons* activate effector cells in the body, such as smooth muscle cells, cardiac muscle cells, or gland cells (figure 15.1). So these postganglionic neurons resemble motor neurons, except that they are located outside the spinal cord and don't control skeletal muscle, but other types of muscle, and secretory glands.

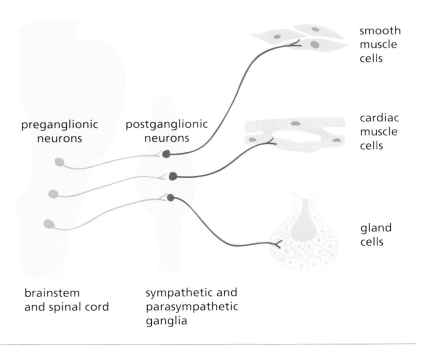

15.1 The **autonomic nervous system** relays brain and spinal cord commands to activate non-skeletal muscle cells and glands through an intermediate set of neurons located in ganglia throughout the body.

The autonomic system is divided into two: the *sympathetic* and the *parasympathetic* systems, which have a different anatomy and essentially opposite functions. Here's the anatomical plan (figure 15.2): the sympathetic ganglia are located in rows parallel to the spinal cord and innervate essentially every organ in the body and the blood vessels that irrigate them. Parasympathetic ganglia, on the other hand, are located right next to the organs that they innervate, and they avoid the skin and skeletal muscle. There are lots of ganglia, each of them of course with a different name, to make things seem complicated and generate nice little questions for the final exam. But they are not complicated; just carefully inspect the diagram in figure 15.2, going slowly from the head to the genitals, and notice that pretty much each organ or part of the body receives dual sympathetic and parasympathetic innervation.

Sympathetic and parasympathetic innervation have opposite functions, due to the different neurotransmitters they release: sympathetic axons mostly release norepinephrine (noradrenaline), whereas parasympathetic axons mostly release acetylcholine. And these transmitters, acting through adrenergic or muscarinic receptors, together with peptides like vasointestinal peptide (VIP), have opposite physiological effects either engaging or disengaging the end organ.

What, then, is their function? The idea is that the sympathetic system is specialized to generate and orchestrate "fight and flight" (or "fight and fright") responses, whereas the parasympathetic coordinates "rest and digest" responses. Essentially, the sympathetic system does action and the parasympathetic does rest. Those two types of responses are used in critical, life-threatening moments in the life of an animal when it needs to either fight an adversary and escape a predator or, on the other hand, recover its reserves when the crisis is over. No wonder this is an ancient structure.

Now, to be efficient at this game, you need to coordinate many different parts of the body. For example, if you are fighting, you need to increase heart rate and blood pressure to better oxygenate the muscles, stop digestion and blood circulation to nonessential organs, inhibit salivation (since you are not up to munching now), widen your pupils to gather more light, widen your bronchia to get more oxygen, stimulate your liver to deliver

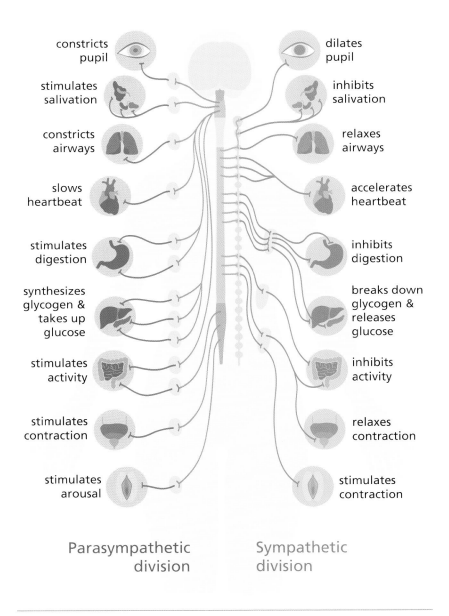

15.2 The autonomic system has **sympathetic and parasympathetic** divisions involved in antagonistic responses to stressful situations.

more glucose, and raise the hairs in your skin to help increase sweating to dissipate more heat. Essentially, the sympathetic nervous system is getting you ready for the fight and optimizing every part of your body accordingly. This is all done in an orchestrated way. Meanwhile, if you want to rest and digest—to recover after a fight and relax—all these organs need to do pretty much the opposite, as well as reengage and stimulate digestion to generate energy to replenish the lost reserves.

This sounds like a nice tight story, but the autonomic system is involved in more things, and this fight versus rest response is not exactly a black-and-white system that you engage when you are fighting or running away. In reality, the autonomic system is much more sophisticated, even though it's still mostly antagonistic, with the sympathetic branch involved in arousal, defense, and escape behaviors and the parasympathetic in eating and procreation. In fact, the anatomy of the system is quite sophisticated, and a sophisticated anatomical structure always means a sophisticated functional plan.

Baroreceptor Reflex

So now you understand why we consider the autonomic system as part of the motor system: at the end of the day, it either activates or inactivates a body response, which is either motor or semimotor. If we continue to think of the autonomic system as a motor system, we see that, like the spinal cord, it also generates reflexes. But whereas the spinal cord innervates skeletal muscle, the autonomic system innervates smooth and cardiac muscle. One example of this is the baroreceptor reflex, which critically controls our blood pressure, raising or lowering it as our posture changes, and controlling the redistribution of blood. This reflex is on all the time. For example, you may have felt lightheaded after standing up when getting out of bed. This happens because every time you stand up, the blood pressure in your head is dramatically reduced, as the blood falls down to your legs due to gravity. Plain physics. Without the baroreceptor reflex kicking in, we would faint every time we stood, or our heads would explode from a high blood pressure stroke when we lie down.

How does the baroreceptor reflex work? Baroreceptors (*baro* is Greek for "pressure") are located in the carotid sinus (in the cranial artery) and in the aortic arteries. They continuously monitor blood pressure using mechanically gated stretch receptors, similar to the ones we have encountered in the somatosensory system, that are activated by high blood pressure or inactivated when blood pressure falls below a particular level (figure 15.3). So when you sit or lie down and blood pressure increases above a certain level, the baroreceptors send action potentials to the nucleus of the solitary tract in the brainstem, which activates a series of circuits that engage the parasympathetic system and disengage the sympathetic system, causing dilation of blood vessels. decreased cardiac

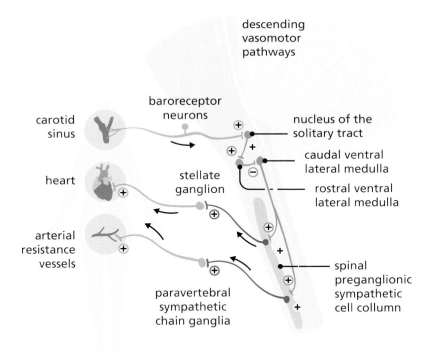

15.3 The **baroreflex** uses a negative feedback to control heart rate and arterial pressure and thus maintain a stable blood pressure during changes in body posture.

output, and decreased blood pressure. On the other hand, when you stand up, your blood pressure drops, the baroreceptors stop firing and stop activating the brainstem, removing the inhibition on the sympathetic system, and that speeds up the heart and contracts the arteries to increase your blood pressure. Your pulse goes up the second you stand up because you are inactivating the baroreceptor reflex. It's true. Try this at home! Measure the pulse in your wrist while you are standing or lying down.

The baroreceptor reflex thus neatly allows you to compensate for the changes of blood pressure in the cranium, ensuring that pressure is maintained, which in turn maintains good oxygen levels. This reflex can be understood as nature acting to fix an engineering problem in the way our bodies are designed, a particularly serious problem in bipedal animals like us. We descended from quadrupeds, and this issue with our blood pressure is part of the price we have to pay for standing upright on our legs.

And the baroreceptor reflex also beautifully illustrates homeostasis, the principles that nature uses to maintain the levels that are necessary to maintain life via reflexes. By the way, this business is all pure control theory, with a set point of ideal blood pressure, a sensor that measures blood pressure, and a negative feedback loop (high blood pressure decreases it, whereas low blood pressure increases it), and a gain that controls the strength of the feedback loop. It's just like the thermostat that regulates the temperature of a lecture hall. This reflex is just one of the few autonomic reflexes that we understand, out of a likely very large number of them.

Enteric System

Let's switch to another part of the peripheral nervous system, one that covers our entire digestive system. It is called the *enteric nervous system* (from the Greek *enteros*, or "intestine"; it's also known as the *visceral nervous system*), and it is essentially disconnected from the rest of the nervous system, to the point that one can argue that it is another brain. Just as if it were the nervous system of a small cnidarian animal, it is built

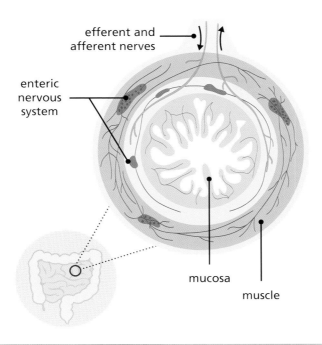

efferent and
afferent nerves

enteric
nervous
system

mucosa

muscle

15.4 The **enteric** nervous system has a nerve net that lines the gut and controls smooth muscle, blood flow, and secretions.

with two parallel nerve nets, or plexuses, in the intestinal wall, one next to the intestinal mucosa and the other next to the muscle layer (figure 15.4). Also, like in cnidarians, the neurons are spread out, mostly without forming ganglia, and are loaded with peptides. What do they do? They control secretions from the mucosa and vasodilation or vasoconstriction of the blood supply, and they also activate or inhibit the smooth muscle in the intestinal wall. Thus, the enteric nervous system qualifies as a bona fide motor system, generating motor patterns.

In fact, do you remember the central pattern generator (CPG) from the lobster stomatoganglion? Something similar is going on all the time in your enteric nervous system, with a CPG that generates peristalsis (figure 15.5). Think of your digestive tract as essentially a tube lined with smooth muscle whose goal is propelling food through the tract. This is

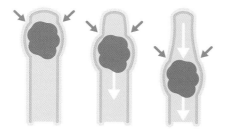

15.5 Peristalsis, the propulsion of food pellets through the digestive system, is generated by an enteric CPG.

achieved by periodic waves of contraction behind the bolus of food with an associated relaxation ahead of it. The circuit that is responsible for this action makes sense: a set of excitatory neurons activate the muscle behind the food while, at the same time, activating inhibitory neurons (with GABA or glycine) located directly in front to relax the muscle. All you need is a circuit with asymmetry in the projections of the neurons from front to back, and this anatomical asymmetry generates the functional asymmetry, which is expressed as peristalsis.

Like any good CPG worth its name, peristalsis can happen independently of the rest of the world: the intestine can be removed from an animal and still generate peristalsis, in the same way that the CPGs located internally in the spinal cord in the generation of a walking rhythm continue to operate independently. And just as the CPGs in the spinal cord are controlled by commands from above (e.g., controlling the pace of walking), the same is found in peristalsis; it can be modulated by stimulating particular nuclei in the brainstem (the nucleus ambiguus or the vagus nerve can trigger peristalsis, for example, when we smell something tasty). Yes, there are relatively few connections between the brainstem and your gut—both afferent and efferent. So the enteric nervous system does listen to your needs . . . sometimes.

Hypothalamus and Pituitary

After this detour in the peripheral nervous system, let's go back to the brain and talk about the hypothalamus, a tiny nucleus at the base of the

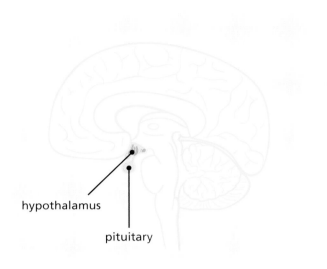

hypothalamus

pituitary

15.6 The **hypothalamus** has a complex set of nuclei that receive inputs from the brain, process them, and activate the pituitary and the brainstem.

forebrain with enormous physiological and clinical importance (figure 15.6). It is composed of many different subnuclei all stuck together. The hypothalamus has a lot of anatomical complexity, but we can simplify it to three types of nuclei: anterior, posterior, and descending. The hypothalamus is connected to a structure called the *pituitary* (or the *hypophysis*), which sits next to the optic chiasm and can be affected by tumors, most of them benign. And, as you might guess. the anterior hypothalamus is connected to the anterior pituitary and the posterior hypothalamus to the posterior pituitary. So far so good.

Now, neurons from the hypothalamus synapse on neurons in the pituitary, which in turn secrete hormones, which are carried out by the bloodstream to the rest of the body. That makes the hypothalamus-pituitary axis "hormone central": it is the place from which the brain controls the hormonal state of the body. This is very important physiologically because the connection between the brain and the body in many ways happens through the hormones released to the blood from the pituitary. As neuroscientists we talk a good game and tend to emphasize the role of synaptic wiring (the *anatomical connectome*), but at the end of the day, it is very likely that a lot of our behavior is controlled by hormonal signals through the bloodstream (the *chemical connectome*).

Which hormones does the hypothalamus secrete and control? Let's go in order from anterior to posterior. In the *anterior axis*, hormone synthesis is a two-step process. First, parvo (small) cells in the anterior hypothalamus synthesize a first set of hormones, called *releasing* factors, that activate receptor cells in the anterior pituitary. These pituitary neurons then release a second set of hormones, which are sent to the rest of the body through the bloodstream. Why have this two-step process? Why can't the hypothalamus just get the job done by secreting hormones into the bloodstream in the first place? We don't know; perhaps it is for amplification purposes, or to enable more flexible regulation, as we have learned in the phototransduction cascade in the retina and other biochemical cascades through the body. Many releasing hormone–secreting hormone pairs use this basic two-step mechanism.

Let's dive into some of these hormones (box 15.1). For example, the thyrotropin-releasing hormone (TRH) binds to cells in the pituitary to trigger the production of thyrotropin (TSH), which then enters the bloodstream, binds to thyroid tissue, and triggers the release of yet a third hormone, which controls the metabolic state of our body and sets the point for the body's metabolism. In the pituitary, TRH also releases prolactin, which controls the production of milk in breast tissue. Another pair is the corticotropin-releasing hormone (CRH), which triggers the release of adrenocorticotropin (ACTH), which then goes to the adrenal glands, where it stimulates the release of cortisol, which seemingly has a billion effects, including the regulation of stress responses. Another pair is growth hormone-releasing hormone (GHRH or GRH), which triggers the release of growth hormone (GH) in the pituitary, which influences bone growth, among other things. Yet another pair is gonadotropin-releasing hormone (GnRH), which releases luteinizing hormone (LH) and

BOX 15.1: Hypothalamic/pituitary hormones

Hypothalamus:

Thyrotropin-releasing hormone (TRH)
Corticotropin-releasing hormone (CRH)
Growth hormone-releasing hormone (GHRH or GRH)
Gonadotropin-releasing hormone (GnRH)
Prolactin release-inhibiting factor (PIH)
Growth hormone release-inhibiting hormone (GIH or GHRIH)
Vasopressin
Oxytocin

Pituitary:

Thyrotropin (TSH)
Prolactin
Adrenocorticotropin (ACTH)
Growth hormone (GH)
Luteinizing hormone (LH)
Follicle-stimulating hormone (FSH)

follicle-stimulating hormone (FSH) in the pituitary, which then trigger the production of steroids in the gonads.

To make it more fun, we also have some inhibiting factors, such as prolactin release-inhibiting factor (PIH), which inhibits the secretion of prolactin by the pituitary, which, as just noted, was stimulated by TRH. In addition, hypothalamic growth hormone release-inhibiting hormone (GIH or GHRIH) inhibits the release of growth hormone (GH) by the pituitary, which was excited by GRH. Thus, for prolactin and growth hormone, there is a push-pull system of antagonistic levers, probably to allow for more precision and control. I'm sure there are similar push-pull control loops of all the other hormones.

How about the *posterior axis* of the hypothalamus/pituitary? Neurons in the posterior hypothalamus (larger, magno cells) project to the posterior pituitary, and their axons branch around the blood vessels (figure 15.7). These blood vessels form a mesh (*anastomose* is the technical term) in order to collect all the hormones synthesized by the posterior hypothalamus and send them out to the rest of the body. There are two main hormones secreted by the posterior pituitary, both derived from the same amino acid backbone: vasopressin and oxytocin. Although synthesized in the hypothalamus, they are released into the blood by the posterior pituitary.

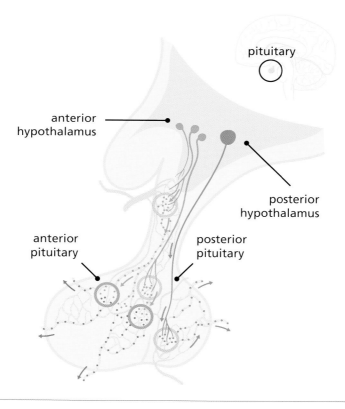

15.7 The **pituitary** is a complex structure that amplifies hypothalamic commands, turning them into hormonal secretions.

Vasopressin (antidiuretic hormone, or ADH) restricts the production of urine (hence the name) and helps to retain water if the blood pressure drops or if the osmolarity of the blood increases. Vasopressin can be considered either a neurotransmitter or a hormone, since the definition of a hormone is a blood-borne substance that has widespread effects throughout the body. Oxytocin is important for uterine contractions during birth and enhances lactation after the birth. It also appears to be responsible for a lot of our social bonding, so it must have widespread and complex actions throughout our brain.

A fascinating story about oxytocin has to do with prairie moles. Yes, studying weird animals like prairie voles is actually relevant for neuroscience. It's a beautiful story: it turns out that there are two subspecies of prairie moles, one of which is monogamous and another which is not, and the difference between the monogamous and nonmonogamous moles seems to be genetically determined by differences in oxytocin expression. Thus, something as apparently behaviorally complex as whether or not you are monogamous may actually depend on your levels of oxytocin, showing how some of our apparently sophisticated behaviors may be predetermined by genetics and put into place through our hormones.

I bet some of you are wondering about the third part of the hypothalamus, the descending hypothalamus. That is the part that talks to the brainstem, influences or triggers autonomic reflexes and CPGs, and is involved in generating emotions. Let's deal with this next.

Emotions

Let's circle back to the earlier question: "What is an emotion?" You can think of an emotion as a reflex (figure 15.8). An emotional stimulus from the outside world is integrated by some emotional control center in the brain, which then triggers the autonomic system to react via the hypothalamus and brainstem. That action engages the three branches of the nervous system (central, autonomic, and enteric) and generates a muscle, autonomic, and enteric physiological response.

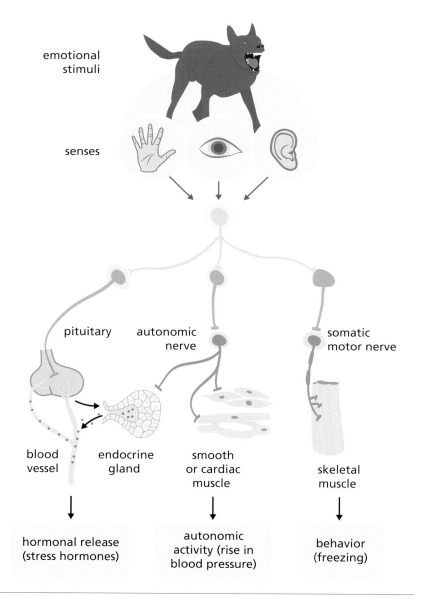

emotional
stimuli

senses

pituitary

autonomic
nerve

somatic
motor nerve

blood
vessel

endocrine
gland

smooth
or cardiac
muscle

skeletal
muscle

hormonal release
(stress hormones)

autonomic
activity (rise in
blood pressure)

behavior
(freezing)

15.8 An **emotional response** works as a reflex, when particularly meaningful sensory stimuli trigger a cognitive response that engages the hypothalamus, pituitary, and the central, autonomic, and enteric nervous systems, and generates behavioral responses.

For example, when an animal feels fear, it freezes, activating the somatic skeletal muscle. In addition, the animal will exhibit an increase in blood pressure as a result of the sympathetic innervation. Moreover, cortisol will be released via the pituitary, again a factor of the autonomic system, and cortisol has lots of effects throughout the body. So, once something is "tagged" by our brain as emotionally charged, a whole slew of downstream pathways are activated. And just like the spinal cord that, in addition to reflexes, also has central pattern generators (CPGs), we have emotional responses at times where there isn't a clear emotional stimulus. This is essentially the reason behind anxiety: we don't have a clear stressor; it is just a consequence of our emotional responses being a type of CPG.

So where is this emotional integrator center that decides whether or not to "tag" a sensory stimulus as emotionally charged and generate an emotional response? Just like the spinal CPGs are not in one particular spot but are distributed throughout the spinal cord (think of it as a feedback neural network), there are many regions in the brainstem involved in emotions. There we find a command center of the autonomic nervous system, sort of the equivalent of the gray matter of the spinal cord. This is called the *central autonomic network* (or CAN), which makes up a major part of the brainstem and is really a big "bag" (or a can) of nuclei that are involved in all kinds of emotional reflexes and CPGs. In the CAN, you connect the CNS, with its conscious processing, with the autonomic and enteric nervous systems. This happens via the CAN's connections to the rest of the brain—in particular, the hypothalamus, amygdala, and limbic lobe, a loose confederation of cortical areas located in the interior of the forebrain (figure 15.9).

The limbic lobe is the part of the cortex involved in the control of many autonomic reflexes, echoing the role played by the motor cortex in modulating spinal reflex and CPGs. The limbic lobe encompasses the cingulate cortex, insular cortex, entorhinal cortex, hippocampus, amygdala, and parts of the temporal lobe. And, just like the cortex is not necessary for walking, it is also not necessary for the expression of some emotions. For example, an animal with no cortex still has emotions. In laboratory

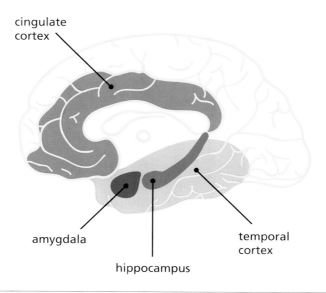

cingulate
cortex

amygdala

hippocampus

temporal
cortex

15.9 The **limbic "lobe"** is a set of cortical areas engaged in emotions.

rats, if you section the forebrain, disconnecting the cortex but leaving the basal forebrain connected to the brainstem, the animal not only can survive (which goes to show you that animals do not actually need the cortex to live) but can actually express rage.

In addition to the CAN, the hypothalamus can also be viewed as an integrator for emotional stimuli. In fact, if the hypothalamus is removed (leaving the brainstem attached), the animal does not have rage, demonstrating that the combination of CAN and hypothalamus is necessary and critical for emotions.

Amygdala

The limbic cortex may not be necessary for expressing some emotions like rage, but it appears to be necessary for emotional learning. This fact resonates with the idea that the cortex is some sort of learning machine. As an example of emotional learning, let's take a look at fear conditioning, which happens in particular in the cortical amygdala (and which has nothing to do with your other amygdala, the one in your mouth). Fear, defined as the response to a threat, is one of the strongest emotions; you learn to associate a neutral stimulus with a particular danger after just one exposure to it, and this can lead to a life-long lesson. One-shot learning.

If you take a mouse and present it with a neutral sound, nothing happens. But if you put the mouse in an electrified grid and play the same sound while you deliver a shock, the mouse will never forget that that particular sound means trouble (figure 15.10). Any time the mouse hears that sound in the future, it will freeze, displaying a stereotypical behavior—a fear response—mediated by the autonomic system.

How does fear conditioning work? It turns out that both auditory (sound) and somatosensory (electric shock pain) stimuli go through the thalamus up into their respective cortices, and in parallel, they also go to the amygdala. All these pathways then converge in the lateral nucleus of the amygdala, which triggers an emotional response. So when the two stimuli (sound and electric shock) happen at the same time, long-term synaptic plasticity (LTP) occurs in the lateral nucleus. And once this

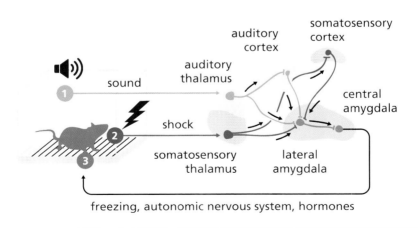

15.10 The **amygdala** associates sensory stimuli that have a strong emotional component, like fear.

happens, future auditory stimuli will automatically activate the amygdala circuits, even without the pain stimulus.

One way to interpret these results is to think of the amygdala as monitoring important sensory inputs and adding to them an emotional tone, according to the emotional salience of the stimulus. If you take out the amygdala, the animals still have normal emotions, but they cannot learn new fear emotions.

And what about humans? In fact, in fMRI scans, when humans are presented with a threatening stimulus, they also engage the amygdala. And, to tie up the story nicely, as shown, patients with amygdala lesions have a defect in fear conditioning. They are not fearless; they just cannot learn fear. These findings all place the amygdala as a generator of emotional behaviors, because we are constantly changing our emotional repertoire. In fact, it seems that both positive and negative emotions are learned by the amygdala, as if it were to give an emotional stamp onto sensory stimuli.

Besides learning to fear new stimuli, some amygdala nuclei are also involved in innate fear, which is preprogrammed in the brain. For example, mice have an innate fear of the urine of foxes. Even if a mouse has never smelled the urine previously and may not even know what a fox is, it will react to the urine with tremendous fear. It is likely that we humans also have innate fears underlying our cognition.

Limbic Cortex

I mentioned that besides the amygdala, other parts of the cortex are involved in emotional responses. As a first pass to this issue, researchers used fMRI while asking the subject to think of or remember positive or negative feelings and searched for the areas that were activated in positive and negative emotions. It turns out many different parts of the cortex light up during all kinds of emotions. We are dealing with a distributed system—again, a neural network, or a network of networks. In particular, there are three cortical areas that get preferentially activated during emotions: the cingulate, prefrontal, and somatosensory cortices.

The *cingulate* cortex, the heart of the limbic cortex, is activated when you are sad or depressed. In fact, patients with clinical depression (major depressive disorder) have a big difference in activity in their cingulate cortex compared to healthy subjects. Because of this, physicians are now trying to stimulate the cingulate cortex as a treatment for severe depression to help prevent suicides. But this is more than just preventing depression: knowledge of these circuits, and their modulation using neurotechnological devices, could lead down the line to the control of emotions through neurotechnology and brain-computer interfaces.

Another area that lights up is the *prefrontal* cortex, which is fascinating but mysterious in its function, as we will discuss in the final lecture. Patients with prefrontal lesions usually show a dramatic change in their personality. They become disinhibited, lose empathy, are emotionally cold, and feel no guilt. A lot of social feelings are associated with the PFC.

The primary *somatosensory* cortex is also engaged during emotions, particularly the nondominant hemisphere. This is surprising. What does emotion have to do with somatosensation? Psychological feelings are complex, whereas the somatosensory system seems so straightforward. However, the fact that the somatosensory cortex receives pain information may be part of this riddle: there is a deep connection between somatosensation and positive and negative emotions. Sometimes it can be appreciated in language. In some languages, often the same words are used for a positive somatosensation and a feeling. For example, "warm" is used for "happy" and "cold" for unhappy. Having a "fuzzy feeling." Being "toasty." In fact, in Basque, a wonderfully expressive language, the word *goxo* means "sweet," "tasty," "pleasant," "comfortable," "pleasurable," all of the above—in other words, that warm feeling of happiness in your belly. The bottom line is that our emotions are anchored in many parts of our cortex, and this is just the beginning of a rich and fascinating field of study.

And, finally, why do we have emotions in the first place? Why are feelings necessary? How about getting rid of the amygdala and eliminating fear forever? Wouldn't we be more efficient animals if we weren't saddled with all these emotions? That might not be such a good idea. It is critical to be able to learn an emotion. Just as it is good to have pain, emotions probably help us keep ourselves safe—they are the guiding light of our behavior, as Helen Keller wrote, our "north star." We behave the way we do because we follow these emotions during our lives. We learn through emotions, both positive and negative. Feelings are the long-term manifestations of our emotions from our cognitive mind—and we use the memory of feelings to imagine future situations to steer our future behavior. This is probably why every animal probably has emotions too.

▼ RECAP

Emotions are complex motor responses that involve all three parts of the nervous system (central, autonomic, and enteric), constitute a guiding compass for behavior, and are also subject to many pathologies. Emotions can be understood as specialized motor patterns with elements of reflexes and CPGs. Although they are mostly unconscious, they can also be triggered by conscious thoughts (feelings).

Emotions and feelings are generated by different cortical areas, loosely termed the limbic lobe, including some areas, like the cingulate cortex and the amygdala, that are specialized for processing emotional stimuli. The limbic cortex is connected to the hypothalamus, a critical nucleus, which, with the pituitary gland, acts as an endocrine and homeostatic integrator, hormonal amplifier, and relay system, regulating body-wide functions such as blood pressure and electrolytes, energy metabolism, body temperature, sleep–wake cycles, and reproductive and defensive behaviors. The hypothalamus acts in concert with the central autonomic network, a series of brainstem nuclei that can engage the enteric and autonomic nervous system, with its sympathetic and parasympathetic branches.

Finally, the autonomic/enteric system, which reaches essentially every part of the body, coordinates body-wide reactions to vital behaviors, like defense or escape in life-threatening situations, arousal in sexual engagement, procreation, urination, defecation, digestion, and peristalsis for replenishment of body resources.

Further Reading

LeDoux, J. E. 2000. "Emotion Circuits in the Brain." *Annual Review of Neuroscience* 23: 155–84. A review by one of the leaders in the field studying how emotions are generated by the nervous system.

Gershon, M. D. 1981. "The Enteric Nervous System." *Annual Review of Neuroscience* 4: 227–72. A fascinating look into our "third brain," which lines our digestive system and can operate independently of the central and peripheral nervous system.

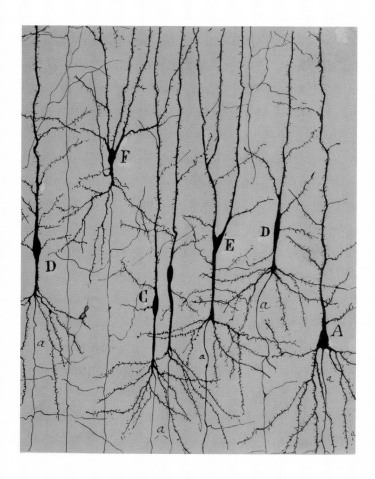

Pyramidal neurons characteristic of the insular acoustic cortex. In a formidable and poorly understood process, the cortex transforms sounds into words and infuses them with semantic meaning and grammar to build language. Courtesy of the Cajal Institute, "Cajal Legacy," Spanish National Research Council (CSIC), Madrid, Spain.

LECTURE 16: SPEECH

The limits of my language means the limits of my world.
—Ludwig Wittgenstein, *Tractatus Logico-Philosophicus* (1922)

OVERVIEW

In this lecture, we'll learn

- ▶ How speech is generated by the brain and vocal tract
- ▶ How speech is decoded by the cortex
- ▶ How neurological syndromes can illuminate principles of brain function
- ▶ How speech develops in infants

Language is the one ultimate human behavior—it is the one thing that, supposedly, other animals cannot do. It's arguably the most sophisticated expression of a sensory-motor transformation in our brain, generating and decoding sounds to generate a symbolic world, narrating our mind. It's the nervous system at its best.

One way to understand the neural basis of language is to treat speech as just another motor output, as a continuation of the lessons of the motor system. But much more of the brain is devoted to sensation and movement than to language, so language is only like a small boat on the immense ocean of the sensory-motor system. In speech, we generate a combination of brief sound-generating motor patterns that serve as building blocks, or modules, of words. Words themselves are also modules of sentences, with which we can create an infinite number of combinations, reflecting an infinite number of concepts and thoughts. It takes years for infants to properly decipher and generate this arbitrary code that links sounds to meanings. And speakers of the same language can understand and decipher that arbitrary code and, through it, share a mental interpretation of the world.

Speech

But what exactly is speech? What does speech look like and how can you measure it? We measure speech with a spectrogram, which is a plot of the sound frequency as a function of time (figure 16.1). If you inspect these plots carefully, you can see that they have a peculiar structure, with areas that represent rapid changes in sound. The basic units are called *formants*, and they represent changes in air pressure generated by different muscles of the larynx. We can generate many different formants, each with a particular pattern of frequency versus time. Some formants

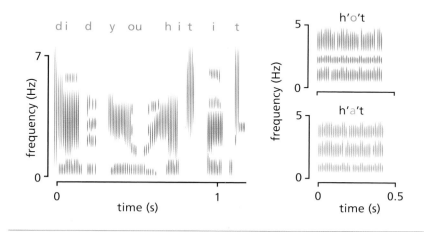

16.1 Speech consists of stereotyped sounds with differences in frequency modulation and harmonics that build vowels, consonants and words.

are stable over time and correspond to vowels, whereas others are brief or change in frequency (that is, they are frequency modulated, or FM) and correspond to consonants. Some formants also have funny-looking ladders of frequencies, which are called *harmonics* and are also critical as building blocks of words.

For example, let's compare the "ah" sound of the word "hot" with the "ae" sound in "hat." Although some basic frequencies are identical in both sounds, the harmonics between "hat" and "hot" are different. This small difference, which lasts only 100 ms, results in different formants, and, in this case, completely changes the word and its meaning. This example demonstrates the exquisite selectivity of the nervous system in its ability to generate and detect these small differences in frequency.

So, when we speak, we generate a continuous stream of sound, and our brains split that sound into formants, assembling words and deciphering their meaning. And all of this works in real time, at the pace of speech. This is amazing because, by looking at the spectrogram, it is difficult to detect where one word begins and another word ends. Detecting words is a huge challenge for speech recognition computer programs, which have great difficulty identifying auditory information and parsing it out. So one insight right off the bat about our speech system is that we are dealing with an incredibly sophisticated machine.

The modularity of speech and language implies that there are several layers of abstraction in the generation and decoding of language, and you can imagine that it could fit well with a hierarchical network model, like Hubel and Wiesel's perceptron. So, to understand language, we would need to first decode sounds, then formants, then words, and, finally, figure out how words map to semantics, the meaning of the words. Linguists define these levels as *phonemes* (sets of sounds) and *morphemes* (semantic units, prefixes, suffixes, etc.), which together generate words, which then are organized in a set of *rules*—grammar—specific to each language. Thus, at its heart, language has a hierarchical architecture, constructing increasingly abstract concepts as building blocks.

Speech Generation

But how are formants sculpted out of pressure waves? It all happens in the lungs, vocal cords, and vocal tract (figure 16.2), and you can think of it as a musical instrument. The lungs provide an initial pressure wave, and the vocal cords generate a series of harmonics of fundamental frequencies of the formants. Then the vocal tract—which includes the larynx, pharynx, soft palate, nasal cavity, and lips—shapes and filters the pressure waves and harmonics.

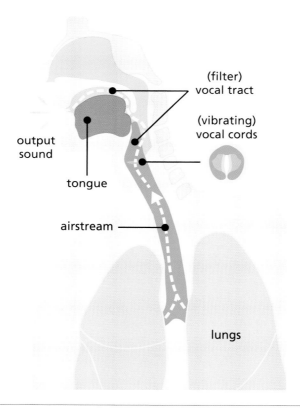

16.2 The lungs, vocal cords, and vocal tract generate pressure waves and shape them into precise speech **formants**.

Superimposed on all of this, another critical part of language is *intonation*—essentially, the way you say things, the nonsemantic content of language. Surprisingly, in human-to-human communication, the content of the speech is only a very small proportion of the actual information being conveyed in a conversation. In fact, the way you say things supposedly carries more information than the words you say. Intonation is also generated and modulated in real time by the vocal system. Additionally, as good actors demonstrate, we transmit a lot of information with our faces and our body language, and that often has a higher communication value than the speech itself. This nonverbal information involves an incredibly complicated and rich set of face muscles, and also our upper limbs, as a gesticulating Spaniard like me can confirm.

Aphasias

To understand how speech is generated and decoded by the brain, let's take a detour and look at a fascinating set of patients with aphasias; that is, our speech deficits, which are often found in stroke patients (figure 16.3). As mentioned earlier, sometimes you can gain deep insights into a system when it malfunctions, and in biology, many breakthroughs have come from studying patients.

Two revolutionary discoveries at the end of the nineteenth century by Paul Broca and Carl Wernicke, astute physicians who studied brain injuries, revealed specific cortical areas responsible for decoding and generating speech. Working in France, Broca noticed a peculiar speech impediment in patients who had a lesion in the left (and only the left) frontal cortex (figure 16.3). When trying to describe a picture, Broca's patients often used the right words, but the way in which the words were strung together was broken. They recognized an object and understood its meaning, but somehow could not convey this information (figure 16.4). Their speech was halting, with repeated words or sentences and disordered syntax and grammar. Although the motor aspect of their speech was altered, the conceptual aspects were intact.

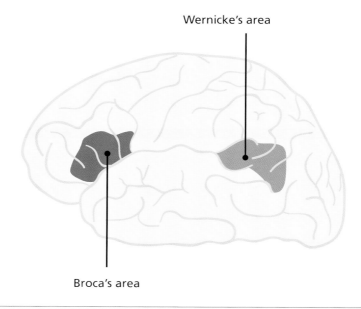

Wernicke's area

Broca's area

16.3 Patients with lesions in **Broca's and Wernicke's** areas have specific deficits in the generation and decoding of speech.

Meanwhile, in Germany, Wernicke noticed that patients with lesions in the upper temporal lobe, next to the angular gyrus, had problems with the interpretation of speech. When asked to describe a picture, they spoke fluently, without repetition and with correct syntax and grammar, but the words were gibberish, not conceptually coherent.

Broca's and Wernicke's aphasias, called *motor aphasias* or *sensory aphasias*, respectively, were not just curiosities, fascinating discoveries about the nature of speech generation; they also provided the first serious evidence for the localization of brain function and, also, for the lateralization of brain activity. These clinical cases proved that there are critical areas for speech decoding in the temporal lobe and for speech generation in the left frontal lobe. And Broca nailed it: speech is lateralized in humans. Even in patients who are left-handed, most speech is generated by the left hemisphere—*"nous parlons avec l'hémisphère gauche"* ("we speak with the left hemisphere"), as he put it. In stroke patients,

16.4 Aphasias are deficits in the generation or decoding of speech.

lesions in the left hemisphere cause problems with speech, but patients are still able to process the emotional and social aspects of speech in their right hemispheres. Meanwhile, strokes on the right hemisphere have the opposite effect: patients can speak fine. However, they speak with very flat intonation. So it seems that the right hemisphere is used for the non-semantic aspect of speech.

Language Processing

From these patient studies, researchers put together a basic model of how the brain generates and decodes language. Language first enters the cortex via audition or reading through the visual or auditory cortex. It then feeds into Wernicke's area, where speech recognition occurs and which charges words with semantic meaning before being sent to Broca's area, which is very close to the mouth and tongue regions of the motor cortex, to generate an output: speech. There are strong connections between Wernicke's and Broca's areas—the so-called arcuate fasciculus, which links them, is a symmetric pathway.

But let's leave these peculiar patients behind and turn our attention back to normal subjects. Where in the brain is language processed? The short answer is: everywhere. If you take a PET or fMRI scan of healthy subjects while they are speaking, many different cortical regions light up—and even regions outside the cortex. This makes sense, because during speech, humans are listening, thinking, generating words, or reading faces or reading texts. Reading written words or sign language, as an example of visual speech, activates the visual cortex, whereas listening to words activates the auditory and upper temporal cortices (consistent with Wernicke's area). The actual speaking activates the upper parietal frontal lobe (consistent with Broca's area), whereas thinking of words activates a wide network of cortical areas. In fact, Broca and Wernicke actually discovered just the tip of the iceberg: there is a larger network of cortical areas involved in language, and Broca's and Wernicke's areas are just nodes of a larger network that includes motor, premotor, somatosensory, temporal, and even visual cortices (figure 16.5).

Although the initial Broca-Wernicke model is not wrong, speech processing is more complex than originally proposed, with many other cortical areas involved in the recognition of words. These are found in the temporal lobe, or the "what" ventral pathway (remember the lecture on vision?). Meanwhile the "where," or dorsal, pathway connects to Broca's area—close to the motor aspect of speech. Speech also ends up involving the striatum, with its strong relation to the frontal and motor parts of the cortex. So the parallel dorsal-ventral streams reflect the way we think and speak, with a fundamental core split between concepts and action. And, by the way, as well as for standard speech this basic model also applies to sign language, reading, and writing, which shows that evolution is

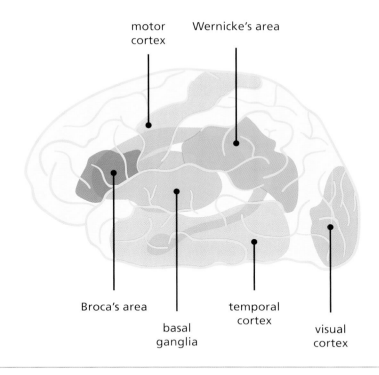

16.5 Language decoding and generation occurs in a wide network of **cortical areas**.

coaxing the communication system to build speech and that humans are tapping into the same circuits as when we invented writing.

Here's a reflection on the history of language research: stimulating or lesioning a brain region to assign a particular function to it, while a good start, may be partly misleading if the region in question is a node within a neural network. You may have tapped into a circuit that could be much larger.

Speech Development

As a beautiful example of a complex behavior, language can also provide deep insight into how behaviors are acquired during development. How is language acquired? Infants are born without language and they learn to speak from their parents and their social environment. If you are observant and have kids, or younger siblings, nieces, or nephews, you can probably reconstruct the process. Already in the first few months of life, babies start to climb up this ladder of increasing levels of language acquisition by listening to adults speaking and first learning the phonetics of the specific language they are being taught. This is followed by babbling, a period in which they start to generate sound, and then another period when they generate combinations of sound that are not specific to the language they are being taught. By the age of one, they are talking. This is a stereotypical progression, identical across cultures and languages.

To explain this process, there are two hypotheses that are interesting not only by themselves but also because they reflect the nature versus nurture debate and tension that exists within developmental neuroscience. The first one, that nurture determines behavior, following the empiricist argument that the mind is a reflection of the world, led to the development of *behaviorism*, represented by B. F. Skinner. He argued that the way babies learn language is via reinforcement; that is, by reward or punishment. He viewed infants as essentially a machine that gets trained by our positive or negative reinforcement of their behavior. We enter the world with a blank state on which all our positive or negative experiences we have of the world are sculpted—a strictly behavioral programming.

The second hypothesis, that nature actually determines behavior, is called *nativism*, defended by Noam Chomsky. This states that language is something innate. This idea reflects a more Kantian view that we already have all these categories inside us. According to Chomsky, we are born with a language "instinct" that expresses itself naturally. With that hypothesis, he solved the problem that behaviorism has: that there is not quite enough time for babies to learn language—it happens so quickly in the first few months of life that they don't really have much time to learn it all by receiving positive and negative rewards.

This nature versus nurture debate is ongoing, and perhaps it's too simplistic, as there are hints of truth in both positions. There is overwhelming evidence for the universality and strict timing of language acquisition across cultures and people, which supports the idea of a preprogrammed

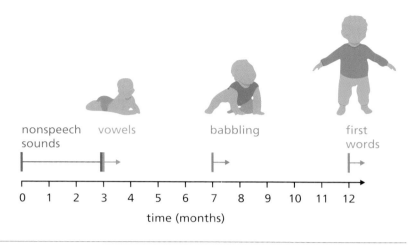

nonspeech sounds | vowels | babbling | first words

| | | | | | | | | | | | | |
0 1 2 3 4 5 6 7 8 9 10 11 12

time (months)

16.6 Language development occurs in a stereotypical progression during childhood.

process (figure 16.6). But, in support of behaviorism, babies are much more sophisticated listeners than we think: they are able to pick up the statistics of speech sounds and even measure Bayesian probabilities of their occurrences. This has been tested by exposing babies to made-up languages and finding that they can pick up those made-up words as easily as those from a proper language. Also, the acquisition of language rules, which happens in the first six to ten months of life, occurs much more efficiently if the baby is reacting with a real person rather than a tape recording or computer-generated speech, demonstrating the strong role of social input in language acquisition. In fact, language acquisition is enhanced by the peculiar (and unavoidable) intonation of adults when they talk to infants. This is called *parentese*: the high-pitched, drawn-out sounds parents make using simplified grammar to make it easier for babies to pick up the sounds.

Language acquisition is also plastic (i.e., capable of change). Any child learning two languages will learn two sets of statistics, but true bilingualism only happens during a critical period for language acquisition. In fact, you can scan the brains of people who are bilingual and determine when they learned the second language (figure 16.7). If you learn two

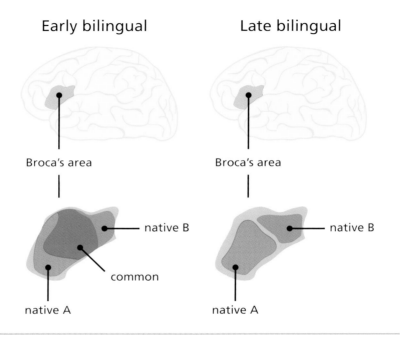

Early bilingual

Late bilingual

Broca's area

Broca's area

native B

native B

common

native A

native A

16.7 Mapping of different languages in the cortex depends on when a **second language** was acquired.

languages early in development, the same parts of Broca's area light up with both languages. So, the same cortical area can process both languages simultaneously. In contrast, if you learn the second language later in life, two different regions within Broca's area light up, one for the first language and another for the second. This finding demonstrates that language learning has a critical period, just like vision. That explains, by the way, why people like myself, who learned English after this critical period, will always carry an accent in the second language no matter how many years we are exposed to English. Fascinating.

Let's put it all together now and make ourselves a model of language development, just like we did for language generation. If you remember the lectures on development and neural networks, you can think of this as a two-step process that occurs in the development of a network. First, a whole bunch of synaptic connections are made very early on in the infant's life—approximating the idea of a neural network that wants to be completely connected. But then a pruning period occurs, which is equivalent to the application of one of the learning rules we talked about in neural networks. The acquisition of speech is equivalent to the building of an attractor landscape. Because of this, in the first nine months, babies' cortical regions are completely plastic; infants can learn any language, from Mandarin to Basque. But after the first year, they experience a fundamental change during which circuits are pruned, and then the baby cannot distinguish the differences between the formants, intonation, and phonetics of different languages. After that, the critical period is over, and the circuit is fixed, and its attractor landscape is set. That's why learning a new language is so tough.

▼ RECAP

Language represents the core of the human condition and is a sophisticated form of communication that uses sound to build a hierarchical architecture of modules—like formants, phonemes (sets of sounds), morphemes (semantic units, prefixes, suffixes), words, sentences, grammar, syntax, and prosody—with which to abstractly represent and manipulate our external and internal worlds. Studies of patients with motor aphasias (broken speech) or temporal aphasias (problems understanding speech, nonsense speech) revealed two cortical areas, Broca's and Wernicke's, which are critically involved in generating speech. But these nodes are part of a larger cortical and subcortical network of areas, which are also activated in sign language and other forms of nonverbal communication, like reading and writing. Language acquisition begins soon after birth and occurs in stereotypical stages, whereby an innate behavioral program is molded by social exposure to speaking adults, reflecting the interplay of nature and nurture in sculpting cortical circuits into effective neural networks, and in which the language could be represented as an attractor landscape.

Further Reading

Chomsky, N. A. 1998. *On Language: Chomsky's Classic Works: Language and Responsibility and Reflections on Language.* New York: New Press. A recollection of Chomsky's seminal ideas.

Hickok, G., and S. Small. 2015. *Neurobiology of Language.* London: Academic. A comprehensive book on the brain mechanisms of language generation and decoding.

Kuhl, P. K. 2010. "Brain Mechanisms in Early Language Acquisition." *Neuron* 67, no. 5: 713–27. A review of the neurobiology of language development.

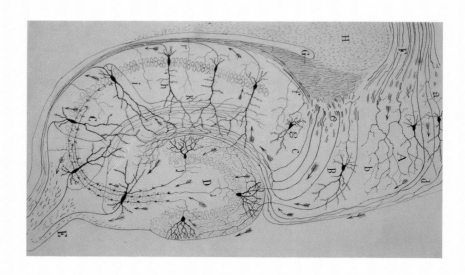

Structure and connections of hippocampus (Ammon's horn). The hippocampus is a beautiful circuit that helps to store long-term memories and could generate our concepts of space and time. Courtesy of the Cajal Institute, "Cajal Legacy," Spanish National Research Council (CSIC), Madrid, Spain.

LECTURE 17: MEMORY

He told me that before that rainy afternoon where he fell from the horse, he was what all Christians are: blind, deaf and memory-less. For nineteen years he had lived as one in a dream: he looked without seeing, listened without hearing, forgetting everything, almost everything. When he fell, he became unconscious; when he came to, the present was almost intolerable in its richness and sharpness, as were his most distant and trivial memories. Somewhat later he learned that he was paralyzed. The fact scarcely interested him. He reasoned (he felt) that his immobility was a minimum price to pay. Now his perception and his memory were infallible. . . . He was, let us not forget, almost incapable of ideas of a general, Platonic sort. Not only was it difficult for him to comprehend that the generic symbol dog embraces so many unlike individuals of diverse size and form; it bothered him that the dog at three fourteen (seen from the side) should have the same name as the dog at three fifteen (seen from the front). . . . With no effort he had learned English, French, Portuguese, and Latin. I suspect, however, that he was not very capable of thought. To think is to forget differences, generalize, make abstractions. In the teeming world of Funes, there were only details, almost immediate in their presence.

—Jorge Luis Borges, "Funes The Memorious," 1944

OVERVIEW

In this lecture, we'll learn

▶ There are two basic types of memories, declarative and procedural
▶ The hippocampus is critical for long-term memory storage
▶ Memories are stored in different parts of the brain
▶ How attractor neural networks can explain memory recall

Learning

In these last two lectures, we are entering relatively unexplored territory: the cognitive and intellectually advanced functions of the brain. Our limited understanding of the nervous system makes it difficult to make rigorous claims about some of the most sophisticated functions of that system, so bear with me. Learning and memory, for example, are two of these sophisticated abilities of the nervous system, and both involve gathering information from the world, storing it, and then recalling it almost effortlessly.

Part of the challenge with studying learning and memory is their definition—what exactly do we mean by learning? Learning a phone number, to drive a car, to walk (but can we really "remember" how to walk)? Francis Crick, whom I was lucky to meet, was not only phenomenally smart but also phenomenally funny. After revolutionizing genetics and molecular biology, he turned his attention to the brain, hoping to understand how it works. When thinking about memory and learning, he was cautious not to get trapped in nomenclature; he broadly defined learning as "any change that makes a change." That was not intended as a joke. Crick believed that anything that changes the function of the nervous system constitutes learning. Such conceptual breadth enables us to avoid early definitions that could straitjacket and launch us into investigating an ill-posed question.

To prevent this misstep, let's stick with Crick's definition of learning and encompass within it any changes in the nervous system that will alter its future function. This all-inclusive definition will also help us avoid the introspection problem: dragging in conceptual and terminology "baggage" about how we think the system should work that could be misleading. It would not be the first time our preconceptions misguide us. Recall how two thousand years of speculation about vision and how we see were put to rest—or, better stated, regrounded—with a single experiment by Hubel and Wiesel in which they discovered that practically all neurons in the primary visual cortex cared about the orientation of edges. Since Hubel and Wiesel, every theory of visual perception starts with edge detection. No one saw that coming. Orientation selectivity demonstrates the danger of using introspection as our guiding light—just because we talk about a concept doesn't necessarily mean that it is real or that this is how nature does it. Nature is independent of our views! So let's keep an open mind when we tackle the brain's more sophisticated functions, cognitive or behavioral, in these final lectures, because what may be buried in our gray matter could be very different from what we think is there.

Memory

So if learning is anything that changes in the nervous system, what is memory? Memory goes hand in hand with learning. We could argue that learning is the process of changing and memory is the change itself. You can also define memory as the encoding by the nervous system of any information. And, it turns out that information can be precisely and quantitatively defined as a reduction in uncertainty. So the acquisition of information can be seen as one of the essential functions of the nervous system: to reduce the uncertainty in our prediction of the future. Yes, this leads us to the conclusion that a lot of what the brain needs to do, to better predict the future, is to store memories. As long as there is any change in the activity of the brain, we can say that the brain has "learned." This idea links us back directly to the principles of neuroscience discussed in lecture 1: it's all about predicting the future.

The study of memory has been very popular in recent years, to the point that one could argue that it has become the center of neuroscience research. And a key part of this is the importance of memory for brain function. As anyone who has a family member or friend suffering from Alzheimer's or dementia knows, memories are essential not just for interacting with the world and for thinking but also for building our personal identities. When we lose our memories, we lose ourselves. How do we recognize that we are the same person in the morning or the evening, or the same person we were when we were four years old, if not by our memory? In fact, memory involves not just our identities but our mental world, our entire world. You could argue that memories are essentially the same as thoughts or, as Kant would argue, as perceptual representations. Perhaps there is no difference in terms of brain mechanism whether we activate a circuit that generates a perception or a thought, or whether we reactivate it internally, as a memory.

The link between memories and thoughts or cognitive states is clear in disease, as Borges's beautiful story explains. Memories are related to concepts, and they form the thread of our mental world. Without the ability to acquire new memories and knowledge, we would only live second to second and without the ability to abstract, generalize, conceptualize, or build any knowledge. I cannot avoid thinking of Kant again, who argued that facts are impressions that we capture from the world using our senses. As mentioned earlier, he also said that until you plug that information into a set framework, it does not constitute true knowledge. Therefore, information can be things we learn about the world, but these impressions only become knowledge when they are assigned a particular value within a set framework. That framework, built with representations, concepts, and memories, is the edifice of our minds.

Now that we are past the definitions and concepts, let's dive into the matter: where in the brain is memory encoded and stored? The answer is pretty much everywhere, because there are many different types of learning and memories, as you can expect given such broad definitions. So what types of memories are there? Depending on whether or not the remembered event is recent, memory has traditionally been differentiated into short-term (or "working" memory, since it is currently in use) or long-term memory, which you access later. Where to draw the line between the two is somewhat arbitrary, with short term being within seconds to minutes and long term, anything longer than that. In fact, some researchers also define an intermediate memory. But this difference is not arbitrary: there is also a fundamental difference between, say, remembering a phone number that you forget a minute later or remembering the name of your first pet or your own phone number, which you might never forget. The basic idea is that short-term memories can become consolidated into long-term memory. Both types of memories can also

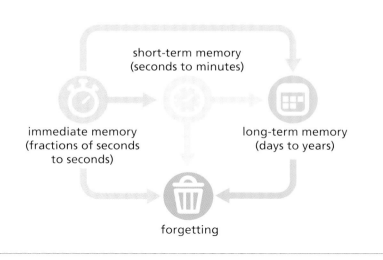

17.1 Information can be stored in **short- or long-term** memory.

disappear into forgetting, another process (or processes), which we understand even less (figure 17.1), and may be necessary for abstraction, as Borges argued.

Short-term Memory

Let's tackle short-term memory first. There are verbal forms of it (i.e., things or concepts that you can describe in words), which, as you can guess, now that you've gained all this neuroscience knowledge, involve storage in the parietal and temporal cortical areas and rehearsal in Broca's motor area. There are also visuospatial forms (i.e., short-term memory of an object you see), which involve the visual, parietal, temporal, frontal, premotor, and prefrontal cortices. As we learned with the what and where streams, there are object and spatial memory subsystems.

From experiments with monkeys, researchers have made some interesting discoveries about working memory—that is, the type of memory you need to operate in the world. In a so-called delay-match-to-sample task, the monkey had to pay attention to an object in a particular position on the screen (figure 17.2). That was followed by a short delay, after which

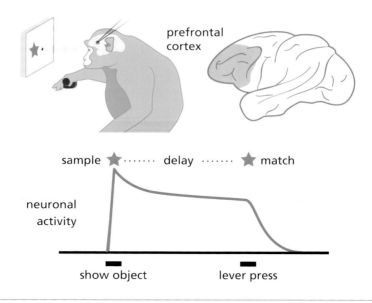

17.2 In a **delay-match-to-sample** task, neurons in prefrontal cortex that code for an object are activated during the period when the object is no longer present but has to be remembered.

the monkey was shown a selection of images. When the object was shown again, the monkey had to press a button to show that it was the object he first saw. He then got a sip of juice, which motivated him to pay attention and remember the sample object.

Well, when you record from the monkey's prefrontal cortex during this task, you find neurons that code for the initial object and are active in the delay period, as if they are holding that image—that memory—in its firing. Interestingly, the neurons' firing returns to its baseline firing as soon as the monkey has pressed the button and no longer needs to remember the stimulus. It is as if the neuron is a switch that the brain has turned on to encode that memory and kept on until the memory is not needed. This switch is on only while the monkey needs to remember the information.

Neurons that fire during delay periods are found in the prefrontal cortex, grouped in territories in which different groups of cells appear to code for the preferred object while others code for the preferred location, and yet others code at the same time for both the preferred object and the preferred location. These findings suggest that visuospatial working memory is encoded in the prefrontal cortex. Of course, we have no method for researching verbal working memory in monkeys, but it could be a similar story as for visual working memory. In fact, many other locations in the cortex contain delayed-firing neurons similar to these: the parietal lobe and the inferotemporal, visual, and premotor cortices. Following the idea that the parietal dorsal stream engages in spatial tasks and the temporal ventral stream is responsible for object recognition, for working memory, we find object neurons, spatial neurons, and some both object and spatial neurons, and all these cortical areas connect reciprocally to the prefrontal cortex.

Models of Memory

How does the brain store information? How can a cortical cell code for a stimulus and maintain the firing in response to a stimulus once that stimulus has disappeared? There are three potential mechanisms, not necessarily incompatible with one another, that can explain this on-switch behavior of working memory neurons.

The first mechanism is persistent *intrinsic* neuronal activity. The experiment is the following: you record with an electrode from a single cortical neuron in vitro (in a dish) and bathe it with acetylcholine. If you stimulate the neuron to fire, it becomes depolarized and generates a burst of action potentials (APs, or spikes). Now, if you stimulate the cell with a longer stimulus, you see a longer burst of APs. So far, so good. But if you now stimulate it with an even longer stimulus, something different happens, as this seems to activate some type of switch that results in persistent firing even after the stimulus is turned off. This mechanism is intrinsic because it does not require other neurons (figure 17.3 [1]).

This is not magic: the individual neuron generates persistent firing due to a positive feedback loop as a result of the channels within the membrane of the neuron. The hypothesis is that when an action potential is fired and the neuron depolarizes, it opens calcium channels in its

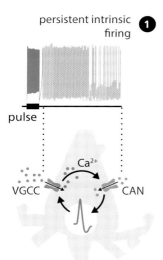

persistent intrinsic **1** firing

pulse

Ca²⁺

VGCC   CAN

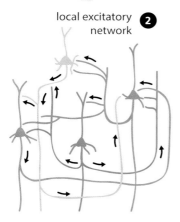

local excitatory **2** network

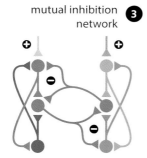

mutual inhibition **3** network

17.3 Short-term memory could be generated by (1) **intrinsic** cellular mechanisms, using a positive feedback loop between ion channels, or by circuit mechanisms due to (2) recurrent excitation or (3) cross-inhibition..

dendrites, allowing the influx of calcium. The intracellular calcium binds and opens cation channels, which further depolarize the cells. And this depolarization brings in even more calcium, opening yet more calcium-activated cation channels. In this way, nature uses two channels that feed off each other to generate a positive feedback loop of activation that acts as a cellular switch.

A second mechanism to generate persistent activity—and, by the way, memory in general—is something you know well: a circuit *attractor* due to excitatory connectivity (figures 17.3 [2] and 17.4). To recap from past lectures, excitatory connections within the neural circuit enable reverberating activity. So a stimulus can have a snowball effect in persisting excitation within the neural circuit. This is the heart of the concept of recurrent neural networks, as we have learned. So, neurons that are active in the delay-match-to-sample task could be connected to others, which connect back to them in an excitatory loop that generates a reverberating activity—persistent endogenous activity. This is a network attractor.

How does an attractor network work, and why is it relevant for memory? We will now retell the story, but from the standpoint of memory. Imagine that we stimulate several neurons to make them fire at the same time, and our goal is to build a memory. Let's now throw synaptic plasticity into the mix so that the neurons that fire together end up being wired together (Hebb's rule). If the experiment is repeated over and over, eventually you wind up with a circuit of neurons that are firing together—as they are interconnected. After these connections are formed, you have an ensemble, which you can think of as a memory. From now on, if you stimulate only one or two of the components of the circuit with a very brief pulse, because they are all preferentially connected, they will generate a reverberating excitatory activity that flows within the interconnected neurons, allowing the neuron to continue to fire even in the absence of the stimulus.

Now let's model this as an attractor. The idea is to visualize all the possible memories you have as a topographic map, with each memory being a valley. Now imagine that you place a ball at the top of a mountain, so the ball will roll down into the valley. This happens because it has potential

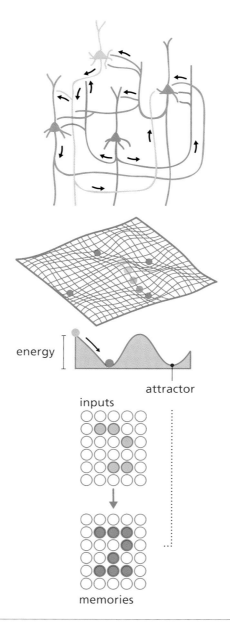

energy

attractor

inputs

memories

17.4 Circuit **attractors**, generated by recurrent excitatory connections, are stable points in the dynamics of a neural network that have pattern completion and can be used to store and recall memories.

energy at the top of the mountain, which is transformed into kinetic energy, which brings the system to the state of the lowest energy. It's just plain physics, Newtonian mechanics. The network of neurons is remembering a particular state that it built previously, because when we trip one neuron, we get the whole pattern—that is called *pattern completion*.

That process is typical of human memory. We often initially remember only part of a thing, but that one part triggers the entire memory. The classic example is from the opening of Proust's long novel, *In Search of Lost Time*, in which a taste memory brings back years of recollections, as I mentioned in lecture 8. In fact, although we are still talking about working memory, this model can be applied to all different types of memory in our brain: every point on this map represents a memory in our brain. In reality, the network map is not in 3-D but in more dimensions—as many dimensions as there are neurons, which is almost impossible to imagine, so a 3-D model is used. From now on, you can think of memories as an attractor landscape.

What about forgetting? This happens when the attractor is erased. For example, the connections in the neural network have been destroyed with metaphorical scissors or a metaphorical filler (like all the information overload we suffer nowadays) that flattens the valley so that activity flows through the area but does not remain there, where the attractor used to be. Incidentally, much of what we have said about attractors applies to the concept of an *engram*, a word used to refer to the physical substrate of a given memory in the brain.

A final mechanism to generate persistent activity to explain short-term memory is a *flip-flop circuit* (figures 17.3 [3] and 17.5) with two excitatory neurons, A and B, connected to a neuron C. Now, neurons A and B are also connected to two inhibitory interneurons—let's call them A_i and B_i—which inhibit the other side. Thus, we have cross-inhibition, analogous to the flip-flop half-center circuit in the spinal cord proposed by Thomas Graham Brown. How does that work?

Imagine that excitatory neuron A on the left is turned on, so inhibitory neuron A_i on the left is activated by neuron A and inhibits excitatory neuron B on the right. So now only the left side fires. But then the left neuron

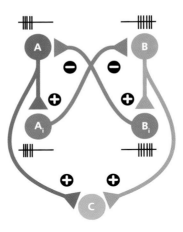

17.5 A **flip-flop** circuit with reciprocal inhibition can generate persistent neuronal activity.

A stops firing (maybe it adapts or runs out of neurotransmitter), so the right side is now disinhibited, causing neuron B to fire, and subsequently inhibits the left A neuron via neuron B_i. The activity alternates between the two sides, but because neurons A and B both connect to C, the result is that neuron C continues to fire indefinitely.

Persistent activity via mutual disinhibition has been found in the oculomotor system: in the performance of saccades, we have an integrator that delivers coordinates to the eye muscles to keep the eye in a particular position. For this position to be maintained, the motor neurons must fire constantly, and they do this through mutual disinhibition. Incidentally, this circuit is mathematically homologous to the attractor model. Although the original Hopfield attractor model was only excitatory, attractors can also be built with these small disinhibitory circuits. In fact, adding inhibitory connections allows attractors to organize different inputs into a topographic map. People use this theory to describe how the cortex can generate *self-organizing maps*, which is fascinating because, as you well know, the cortex is full of maps, and this may be how they come about.

The whole point of this lecture is to try to explain the neural basis at the heart of memory—how memories have some sort of switch that, when turned on, enables us to remember something. These three models seek to explain how that could be built in the brain.

Long-term Memory

So now that we have stuffed ourselves silly with theoretical models, let's come back to earth with a clinical case—one that beautifully illustrates how the brain specifically builds long-term memory. This is another of those Hubel and Wiesel or Broca and Wernicke-type moments, where a deep insight into how the system works almost accidentally drops from heaven and shows us the way forward. As we just discussed, such insights often occur with clinical cases when astute physicians who are studying the symptoms of a patient generate a hypothesis to understand how the system works normally.

The patient in this case, who was called H. M., was a bright child who had a bicycle accident when he was seven years old, in which he hit his head on the curb. From then on, he had persistent seizures that did not respond to medication, with a huge negative impact on his life. Back then—and to this day—the only cure for intractable seizures, those that don't respond to pharmacological treatments, was neurosurgery: surgeons need to extirpate the part of the brain generating the seizures because there is no way to stop them. So, at age twenty-seven, H. M. went to a neurosurgeon, who removed the region of the brain that caused the seizures: a big part of his entorhinal cortex, subiculum, and hippocampus on both sides (figure 17.6).

The operation was a complete success, since all his seizures were gone. But after the surgery, H. M. was left with an enormous memory deficit. He remembered everything before his operation but was unable to recall anything that occurred after his operation. This condition is called *anterograde amnesia*. He would always treat his doctors as if he were meeting them for the first time. He was always living in the present. But this memory problem was quite peculiar, as his neuropsychologist, Brenda Milner, noticed, because he had preserved intact another type of memory, called *procedural memory*—the type of memory associated with skills,

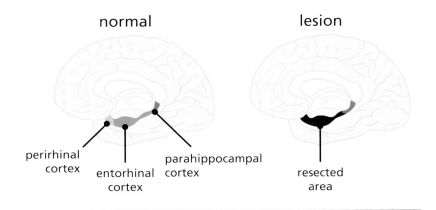

normal   lesion

perirhinal cortex   entorhinal cortex   parahippocampal cortex   resected area

17.6 The hippocampus was surgically removed in **patient H.M.**, leaving him with anterograde amnesia for any new events.

enabling him to perform certain tasks. For example, H. M.'s performance at tracing a star while viewing his hand in a mirror improved over time (figure 17.7), but on subsequent days he had no recollection of ever having done the task. Somehow his brain and body remembered the task, but his mind and consciousness did not. Procedural memory is also known as *implicit memory*. Further tests demonstrated that in addition to normal procedural memory, H. M. had normal working memory, normal to high IQ, normal language, and normal long-term memory of events before the operation. After surgery, he had zero long-term memory, with full-blown anterograde amnesia.

In short, H. M. had a very specific problem with his memory for events, a type of memory called *declarative* (because you can retell the events) or *episodic* (because it is based on "episodes") memory (figure 17.8). He somehow could not transfer declarative information from short- to long-term memory. Thus, the case of H. M. tells us loud and clear that there are two types of long-term memory: episodic (declarative) and implicit (nondeclarative), and that the hippocampus is necessary for the long-term consolidation of episodic memory. In H. M., the areas extirpated were the amygdala, hippocampus, subiculum, and the parahippocampal gyrus at the heart of the temporal lobe. By comparing his case with those

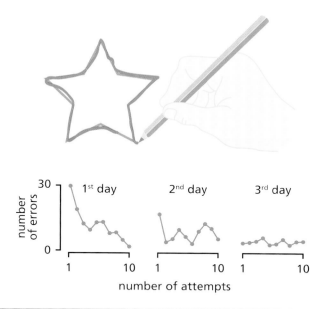

17.7 Patient H.M. had normal **procedural memory**, as his tracing skills normally improved with practice.

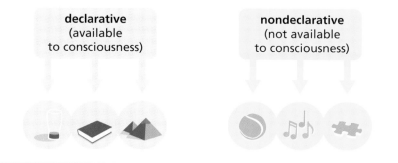

17.8 Declarative memory refers to memories of events that can be stated, whereas non declarative memories are mainly automatic tasks.

of other patients who also exhibited problems with long-term explicit memory as a result of small lesions in the hippocampus, neuroscientists were able to show that the process of consolidation—or transfer—of episodic memory occurs in the hippocampal formation.

Hippocampus

So what is going on in the hippocampus? The hippocampal formation is under the cortex and is called the *hippocampus* (Greek for "sea horse," because it sort of looks like one). This area is also known as the *archicortex* (evolutionarily, the oldest part of the cortex), as opposed to the *neocortex* (the newer cortex, including the occipital, parietal, temporal, and frontal cortices), and the *paleocortex* (the less ancient cortex, which involves the entorhinal and perirhinal cortices).

It turns out that the hippocampus receives connections and connects to a large part of the cortex, so it is ideally suited for consolidating memories and transferring them to the cortex (figure 17.9). Moreover, the core hippocampal circuit is particularly prone to long-term synaptic plasticity (LTP), so it seems ideally suited for the job of taking information from

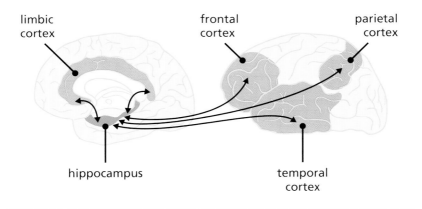

17.9 The **hippocampus** is reciprocally connected to many cortical areas and sends them memories for their consolidation.

all over the brain, building associative memories between different types of information, and sending this message back. And guess what? LTP is generated by repetition. This is just what you would need to consolidate short-term into long-term memory.

After H. M., patient studies continued to reveal the role of the hippocampus and other parts of the brain in memory formation. For most patients, storage of spatial declarative information is affected in right hippocampus lesions, and storage of verbal, or semantic, declarative information in left hippocampal lesions. Isn't that interesting? Space on the right and language on the left. But notice that H. M. had normal long-term memories of events before the surgery. So once the memories were consolidated, they remained intact somewhere else. The idea is that the final storage of all memories is in the neocortex, striatum, cerebellum, and even in the spinal cord and brainstem (figure 17.10). And a

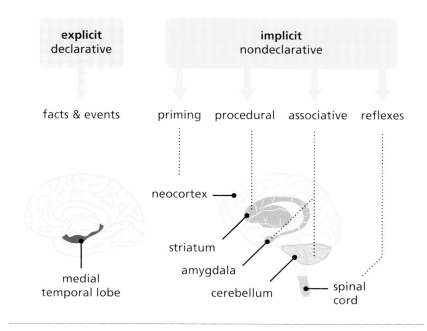

17.10 Different types of **long-term memory** are stored through the central nervous system.

lot of memory storage involves transferring from declarative to implicit memories, as if our brain wants our cognitive mind (or our cortex) to be uncluttered. Just like when you begin to learn how to drive a car, you have to think about every movement, but after a lot of repetition, driving becomes automatic.

In fact, from careful studies of clinical syndromes, you can pin down specific types of long-term memories to specific brain regions in a way that sort of makes sense. The more basic reflexive memories, formed by habituation and sensitization, are found in the spinal cord and brainstem. More complex types of memories, formed by associative learning such as classical or operant conditioning, are found in the cerebellum (motor memories) or amygdala (emotional ones). Memories for complex procedures, skills, and habits involve the striatum, and memories of events and objects, as well as sensory priming, the cortex. Priming is when previous exposure to a stimulus prepares you for subsequent exposure, making you more sensitive to future occurrences.

So, our whole brain could end up being a giant storage container of memories, a giant attractor map. And when one of these attractors is reactivated, you recall a memory. In fact, fMRI studies have shown that recalling a memory reactivates the same regions involved in memory acquisition. Thus, memory is a constructive process—when recalling a memory, you are essentially making up the memory again. That's why these types of studies have had a big impact on legal cases, because in many instances what we remember is not necessarily what actually happened. Memory is always a re-creation, and recall is a constructive, active process. Kant is probably smiling, again.

Space and Time

Before we end the lecture, let me mention something else that is going on in the hippocampus. This fascinating story begins when John O'Keefe and Lynn Nadel discovered by chance (again) that neurons in rats' hippocampus fired when the animal was in a particular position in space! And different neurons fired when the animal was in different positions. They termed these neurons *place cells* and argued that the cells' receptive field was a specific point in space, as if they coded for a particular position. So, the hippocampus seemed to code for space, a map of the world.

Following this, another discovery (also by chance), by Edvard and May Britt Moser, revealed that neurons in the entorhinal cortex fired when the animal was located at regular positions in space, at the nodes of a spatial grid that extended around the area where the rat was. In other words, a given neuron would fire if the rat was in a particular position (like a place cell) but now, they let the rat run around and found that the same cell would also fire if it moved to another location, but at a precise distance from the original spot, in various directions. When they plotted all the positions where a neuron fired, they found that they fell onto nodes of a regular hexagonal grid! Imagine their surprise. They called this new type of place cells "grid" cells. These "grid cells" also seemed to build a spatial map, but this map was now *allocentric*, meaning that it existed independently of the position of the animal, as opposed to the *egocentric* map built by hippocampal place cells, in which the map is always anchored by the position of the animal. From these studies, the hippocampus and entorhinal cortex, which are close to each other and reciprocally connected, appear to be a cornerstone of how mammalian brains internally build space.

What amazes me is that the grid cell map is mathematically regular, with the positions in which the neurons fire located at precise, regular intervals in space, as if our brain internalizes Euclidian geometry. This is pure Kant, as he argued that space and time are internally constructed and superimposed onto the world. That's why mathematics may work, because of a priori knowledge; that is, it is internally based.

Following these discoveries, recent studies are pointing out that place cells also code for time, specifically firing at a particular time during a behavioral task. It's still early in the game, but it sounds as if the space-time continuum that forms the fabric of our world could have to do with the hippocampus. Isn't it fascinating? Understanding our brain mechanisms could reveal the logic of our minds.

Now there is a glitch here, as some of you are probably realizing. Didn't I just tell you that the hippocampus is involved in memory consolidation? So what is this now with space and time? Was this a typo? Are we still talking about the hippocampus? Are there two different parts of the hippocampus, each with a different function? How confusing! Let's try to reconcile it all: the hippocampus could be doing both things, storing memories and encoding space-time. An intriguing idea is that perhaps both functions are the result of the same circuit. Let me explain myself. (This is a hypothesis, by the way.) Perhaps the hippocampus originally evolved to encode space and time, but the same hardware was then used to encode, and anchor into the brain, symbolic "spaces," representing concepts, ideas, or memories. The idea is that, when you think, you think you are mentally walking from one idea to another, as if you are moving from one place to another in physical space, except that now this is all mental space.

Indeed, mnemonists, people with phenomenal memory skills, have known since ancient times that a good trick for remembering a long list of things is to place them, one by one, into an imagined physical map. It's possible to think there could be a fundamental similarity between coding for space and for memories and that, for example, episodic memories could correspond to egocentric maps of our mental world, whereas semantic memories, which are more abstract, could correspond to mental allocentric maps. This all would fit well with the hypothesis that the evolution of the nervous system corresponds to the encephalization and abstract symbolic use of CPGs that generate fixed motor patterns. I find all of this really exciting, as we can at least begin to imagine how studying these weird-looking brain regions with funny names can give us a glimpse of how space and time are created by our brains and how this same exact machinery is repurposed by evolution to build our minds.

▼ RECAP

Learning and memory are broad concepts that likely encompass many different brain mechanisms whenever neural circuits throughout the CNS change after an event. Memories are the building blocks of our identities and mental world and are critical for any cognition. Information about the probability of occurrence of past events, stored as memories, is also a key part of any predictive model of the future. Memory can be short term and long term; there are also declarative memories for concepts or episodes and implicit memories for more automatic motor patterns.

To maintain memories in time, neural circuits have different mechanisms to generate persistent activity patterns, and the prevalence of recurrent excitatory connections in the brain could serve to implement attractor maps of memories or concepts. Whereas the hippocampus is necessary for the consolidation of declarative types of memories, their storage involves the cerebral cortex. Nondeclarative memories of many different types are locally generated and stored by different parts of the nervous system, from the spinal cord to the brainstem, cerebellum striatum, and all throughout the cortex. In fact, one can think of the entirety of the nervous system as a gigantic learning machine that stores events and discretizes them into a continuous tape of space and time, which is internally generated by our brain. This tape of events in space and time is the fabric of our lives.

Further Reading

Hafting T., M. Fyhn, S. Molden, M. B. Moser, and E. I. Moser. 2005. "Microstructure of a Spatial Map in the Entorhinal Cortex." *Nature* 436: 801–6. The amazing discovery that the entorhinal cortex contains a directionally oriented, topographically organized neural map of space.

Kandel, E. R. 2007. *In Search of Memory: The Emergence of a New Science of Mind*. New York: Norton. An interesting autobiography of one of the pioneers of the study of memory.

Liu, X., S. Ramirez, P. Pang, et al. 2012. "Optogenetic Stimulation of a Hippocampal Engram Activates Fear Memory Recall." *Nature* 484: 381–85. Amazing but true: manipulating memories optically in mice.

Luria, A. R. 1987. *The Mind of a Mnemonist: A Little Book about a Vast Memory*, rev. ed. Cambridge, MA: Harvard University Press. The tale of a person with an extraordinary memory.

O'Keefe, J., and L. Nadel. 1978. *The Hippocampus as a Cognitive Map*. Oxford: Oxford University Press. A landmark book that anchored the idea that the hippocampus forms maps of the world, launching a neuroscience approach into a core problem in philosophy.

Sacks, O. 1999. *Awakenings*. New York: Vintage. Another set of fascinating clinical cases by Sacks.

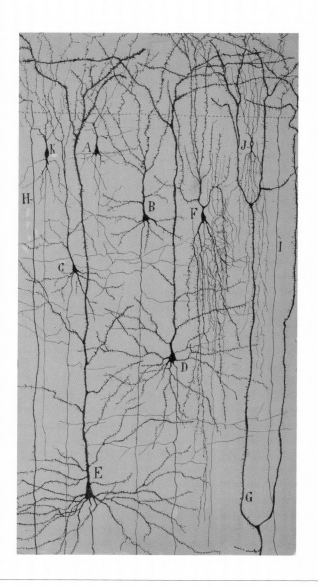

Frontal cortex circumvolution. The neural circuits of the frontal cortex are responsible for our cognition and build a mental world that includes a representation of our own self. Courtesy of the Cajal Institute, "Cajal Legacy," Spanish National Research Council (CSIC), Madrid, Spain.

LECTURE 18: THINKING

All experience is preceded by the mind, led by the mind, made by the mind.

—Buddha, *Dhammapada*

OVERVIEW

In this lecture, we'll learn

- ▶ How association cortical areas link perception with action
- ▶ How attention may work
- ▶ How the prefrontal cortex enables higher behavioral and cognitive abilities
- ▶ What sleep is and what it could be good for
- ▶ How consciousness could arise

Association Cortex

OK, here we are, at the last lecture, aiming to understand one of the most exciting yet difficult problems in neuroscience: the neural basis of thinking. Let's revisit the argument that the brain is some sort of predictor of the future and that it does so by generating a virtual reality model in which the world can be mentally manipulated. The brain then uses this model to choose an action among many, canceling all the other ones, and updates the model based on the outcome using reward prediction or similar mechanisms, and stores this information into memories.

Roughly speaking, you can call all this mental manipulation of the world *thinking*. But in reality, we are talking about a lot of things, about all kinds of cognitive processes. These are complex computations, so this lecture should be taken with a grain of salt because, given our general ignorance about how the brain works, research on this type of high-level questions is still preliminary. In fact, if you define cognition broadly as the study of the human mind, there are many theories about how the mind works, almost as many theories as there are philosophers and researchers. And that's not good, as they often cancel each other logically.

The idea is that cognition is a property—perhaps *the* central property—of the cerebral cortex. So to understand cognition, we need to understand the cerebral cortex. Traditionally, the human cerebral cortex has been subdivided into forty-six areas, identified based on histological differences (though, as revealed through newer molecular and imaging methods, there are probably closer to a hundred). So cognition could be what results from the interaction of all these cortical areas. How do we put all this mess together?

Alexander Luria, a Russian psychologist, proposed a simple triangular model, in which cognition is what sits between the sensory and motor functions at a higher level (figure 18.1). In this model the cortex generates cognition through a hierarchical process in which the areas at the bottom of the model are primary sensory areas (e.g., the primary visual, primary auditory, and primary somatosensory cortices). These areas then connect, in a hierarchal fashion, to secondary sensory areas. The hierarchy then continues to tertiary areas, where coding becomes more and more

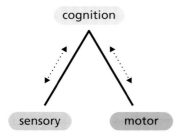

18.1 The **cognitive functions** of the cortex sit between sensory and motor transformations.

abstract. At the top of the triangle, something magical and mysterious happens, at the heart of cognition/heart of the mind. And then these cognitive areas "talk" back down to tertiary motor areas, which in turn talk to secondary motor areas, and primary motor areas and then feed into the spinal cord and generate behavior.

Luria's model, which fits right in with the Hubel and Wiesel model of hierarchical processing, is consistent with the layout of the cerebral cortex and the connectivity among areas. There are clearly sensory areas and also primary and secondary motor areas, so you could argue that cognition happens in all the other areas that are in between. These in-between areas are known as *association cortex* (a term coined by Cajal to describe areas that receive input from multiple sensory areas). The association cortex occupies most of the frontal, parietal, and temporal lobes; also, a significant part of the occipital lobe; and the so-called limbic lobe (Figure 18.2). In fact, it takes up the majority of the cortex in humans. So, to understand the biological underpinnings of cognition let's now be systematic and discuss the functions of each of these association areas, focusing on the temporal, parietal, and frontal lobes, one at a time.

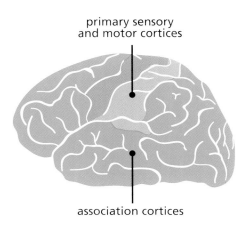

primary sensory
and motor cortices

association cortices

18.2 The **association cortices** (green) are all the cortical areas that are not primary sensory or motor.

But first, let's take a quick detour and talk about the nitty-gritty of brain imaging, which is the primary technique for exploring the function of the association cortex.

Brain Imaging

The study of human cognition has advanced tremendously by the development of methods to image brain activity in humans. The so-called functional imaging methods use different strategies to measure the activity of the brain. Most of these methods rely on tracking metabolic changes in the brain, since we don't yet have methods to directly measure neuronal activity in humans in ways that are not invasive. Why do we care about neuronal metabolism? It's estimated that the brain uses up to 30 percent of all the energy of the body, and most of that energy, in the form of ATP, is used to maintain the membrane potential by feeding the membrane's ionic exchangers. So if a neuron fires a lot, it uses up its ATP quickly, and to replenish it, there are significant increases in the amount of glucose and O_2 in the environment around it, as the neuron uses up both glucose and O_2 faster than usual. Metabolic indexes like glucose or O_2 are thus correlated with the firing of the neuron.

One of these functional brain-imaging methods is called *positron emission tomography* (PET), which takes advantage of the fact that you can image radioactive glucose. PET uses scanners with an array of gamma ray detectors and triangulates the position of the radioactive glucose and thus measures in time and space glucose concentration and its metabolism.

But the most commonly used method is called *functional magnetic resonance imaging* (fMRI), one version of which is *blood oxygenation level dependent imaging* (BOLD), which magnetically tracks the concentration of oxyhemoglobin, which itself reflects the O_2 used by the brain. It turns out that when there is a lot of neural activity there are homeostatic mechanisms that infuse the brain with more blood. This could be part of the job that astrocytes are doing. Because there is more fresh new blood, the oxyhemoglobin is diluted, so you end up with a decrease in oxyhemoglobin, The bottom line is that if you measure oxyhemoglobin, you can indirectly track brain activity. And since you scan each voxel of the brain in parallel with a magnet, you can build an image of the brain in action (figure 18.3). And this is noninvasive! No injections of radioactive glucose, just scanning your own natural oxyhemoglobin. Brilliant! So we can take an fMRI of a person performing a cognitive task and pinpoint the part of the cortex involved, as you will see in the rest of the lecture.

As exciting as this type of brain imaging is—and I can tell you first hand, as I was present at Bell Labs when Seiji Ogawa, Kamil Ugurbil, and David Tank obtained the first BOLD images from rats—fMRI also has limitations. Again, it measures oxyhemoglobin, which only indirectly reflects neuronal activity, and it also has relatively poor spatial and temporal resolution. Keep this in mind, since it is relatively easy to obtain fMRI images (as long as you have a scanner), and they look quite spectacular. Unfortunately, some researchers tend to overinterpret them and fail to perform rigorous control experiments. So keep those limitations in mind and be cautious when reading the fMRI literature.

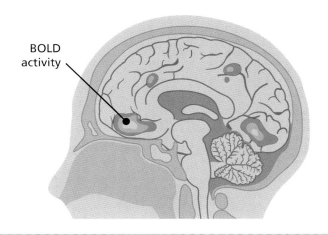

18.3 Functional magnetic resonance imaging (**fMRI**) measures changes in oxygen levels in the blood.

Temporal Cortex

Let's start picking away the association cortex by looking at the cortical areas in the temporal lobe. As I have been drilling into you throughout the lectures, we process all of our sensory information in our different primary sensory areas and then channel this information into two streams, the dorsal and ventral ones (figure 18.4). I covered this in the lecture on vision, but now let's go a little deeper.

Nature divides the two jobs (location and identification) by organizing information into these streams, ending up in two parts of the brain. By doing this, we can compute details in one stream independent of the other. For example, in the ventral stream, we can think about an object without having to know exactly where it is. And vice versa: we can think about space without actually knowing what is in that space. The temporal lobe is part of the ventral stream, receiving inputs from the visual areas

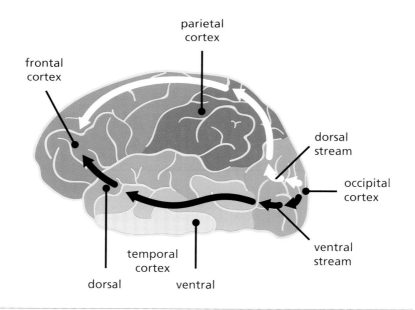

18.4 The **dorsal and ventral** streams process separately information about spatial location and object identity.

in the occipital lobe and projecting mostly to the frontal lobe, the amygdala, and the hippocampus. So, the temporal lobe essentially decomposes the world into a series of objects—a gigantic cabinet, as the philosopher Locke would argue, that starts out empty and is filled drawer by drawer as we live and gain more experiences. Interestingly, the ventral stream is also linked to our emotional response to an object. We cannot see an object and put it in its cabinet without having some sort of emotional reaction to the object. There is a pathway from the temporal cortex to the frontal cortex that mixes in with information coming in from the amygdala. This means that information uploaded to the frontal cortex already has some emotional attachment to it, which helps to trigger our decision-making, which reverberates around the frontal cortex.

How do we know that the temporal cortex is involved in the identification of objects? From four types of experiments: functional imaging, electrical recordings, electrical stimulation, and lesions. Let's talk recordings and lesions. As you may remember from the vision lecture, neurons in the inferotemporal cortex of monkeys fire if a specific object appears in the visual field (figure 18.5). The neuron locks onto the object—basically a grandmother cell—and its firing is independent of any influence (if the monkey is moving or if the object is moving). Other neurons respond to sound patterns. Additionally, there are neurons that code for knowledge—that is a semantic label for the meaning of a particular object; for example,

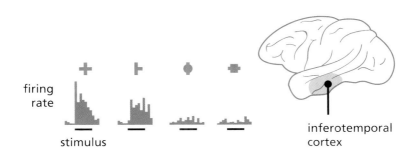

firing rate

stimulus

inferotemporal cortex

18.5 The monkey **inferotemporal** cortex has neurons that respond to specific shapes and objects.

neurons that fire in response to a cross or an object with a pointy shape, demonstrating exquisite specificity.

But maybe the most revealing are the clinical cases of agnosias (i.e., deficits in knowing), a fascinating series of cognitive deficits related to temporal lesions (figure 18.6). In fact, there are patients with object agnosias, in which entire concepts, like vegetables, are wiped out of their minds and they can't recognize them, even though they can describe how they look perfectly well. Related to agnosias are prosopagnosias—face recognition deficits (see Oliver Sacks's book, discussed in lecture 6),

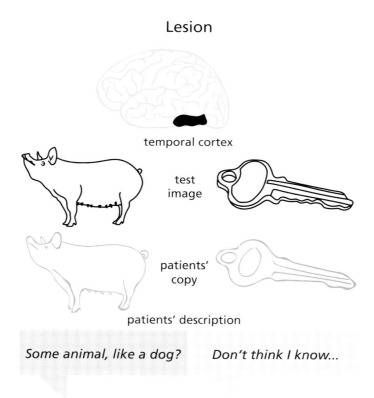

18.6 Patients with **agnosias** have difficulty in recognizing specific categories of objects, even though they can see and draw them perfectly well.

which are probably more common than we think. A significant part of our temporal cortex is devoted to faces, and, indeed, face recognition abilities in humans are extremely sophisticated (we can effortlessly recognize a huge number of faces) and important for our species.

There are also auditory agnosias, where the patient cannot recognize complex sounds. As we saw in the last lecture, in Wernicke's aphasia (speech comprehension), patients have deficits in word recognition; speech generation is garbled, suggesting that you need to have an idea of the word before you are able to use it in language. Finally, there is generalized semantic dementia—in Alzheimer's disease, which generally affects the temporal lobe, where the cabinet empties out little by little over time; patients begin to lose the idea of things, and the world becomes impoverished. If you have a close relative who suffers from dementia, as I do, you painfully watch the gradual disintegration of a mind, bit by bit.

Parietal Cortex

Let's move up to the dorsal stream. A key part of the association cortex is the parietal lobe. This is the heart of the dorsal stream—the "where" or spatial information center of the brain. From electrode recordings in monkeys, which have a parietal cortex that is homologous to that of humans (albeit smaller), we know that it is full of maps, of different types of spatial maps. The general idea is that the parietal cortex has a hierarchy of three types of spatial maps. First maps of the body—skin maps, like the somatosensory homunculus. Then, these personal maps are converted into maps that provide information about the space directly surrounding our bodies—peripersonal space or near-body maps. From here, information is then converted into maps that provide information on global space—the far-body space. (By the way, I find these three types of space categorization fascinating from a linguistic point of view, since Basque has demonstrative adjectives that differentiate between these three types of concepts for space: what you can touch, what is near but not at reach by the hand, and what is far.) The cortex somehow organizes our behavior using these spatial coordinates. The maps are located in different anatomical parts of the dorsal area.

The receptive fields of parietal neurons are quite varied. There are neurons that respond to visual attention, others that respond to head-center maps, others to hand-center maps, and, finally also, neurons that respond to the spatial attributes of the image (figure 18.7). But it is more interesting than that. There are some neurons near the top of the parietal lobe (superior) that have a retina-centered map. For example, when the monkey is looking straight ahead, the neuron fires only if the object falls into the receptive field on its visual field (i.e., uses retinal coordinates). In addition, other neurons respond to head-center coordinates. That means that even if the monkey moves its head the neuron continues to respond, as long as its eyes remain fixed on the target. However, if the monkey moves its eyes, the neuron stops responding. So the map is rotating according to the position of the head. Not bad!

There are also parietal neurons with more complicated responses. In this case, the neurons are not just responding to the particular object the monkey is trying to touch but also to the direction of the reach or trajectory of its hand. Some neurons are specific to the trajectory of a single hand; others respond to the motion of both hands together. This shows that there are different maps for the use of a single hand or when we use both hands. Finally, there are neurons that respond to the grasping of a particular object in a position in space but not to the grasping of a different object in the same position. This is interesting because it shows that the neuron has information about the object, even though it is in the dorsal stream.

As with studies involving the temporal lobe, the study of parietal lobe lesions in human patients has revealed some of the most fascinating insights into its function. Patients display different symptoms based on the location of the lesion. Many of the deficits can be categorized as *neglect* syndromes, in which the patient ignores part of the space, either real space or abstract, conceptual space. If the lesion is on the dorsal portion of the parietal lobe, the patients tend to have motor or "real" spatial symptoms. Some patients have *asomatognosia* ("lack of body

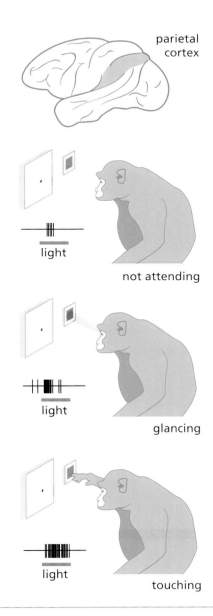

18.7 Neurons in **parietal** cortex can respond to particular positions of an object in space, if they are interesting to the animal.

knowledge"), in which they lose body awareness. For example, with a lesion on the left parietal cortex, you ignore the right side of your body, as if didn't exist. It still has motor and sensory function, but it is somehow out of your mind.

Neglect syndromes can also extend into the space beyond your body, so the patient can also ignore their peripersonal and far-body space. Since lesions are often unilateral (because it is almost impossible to have a stroke in exactly the same place in both hemispheres at the same time), most patients have hemineglect. For example, if you ask the patient to draw an object, they will ignore one side of it (figure 18.8). This is fascinating since the patient still has intact vision of those neglected locations, but somehow they are now out of their awareness.

To make it more interesting, parietal patients often suffer from *anosognosia*, the denial of neglect. For example, some patients complain that there is an alien right arm in their bed while they are sleeping at night. Frequently, even when shown that their arm is connected to the rest of their body, the patient still denies that the arm belongs to them.

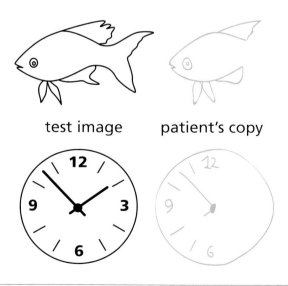

test image   patient's copy

18.8 Hemineglect patients ignore one half of the world.

Other dorsal parietal lesions generate *ideomotor apraxia* (lack of skilled movement), as if the messing up of spatial maps results in unskilled motor behavior, which would be expected. For example, a patient is asked to put their hand in a slot but is unable to do so, due to misorientation of the hand, even though the patient has complete normal motility of the hand. This condition demonstrates a problem with spatial coordinates that the motor system is trying to use, which are somehow abnormal. Another condition is *optic apraxia*, involving problems with visual behavior. For example, the patient is asked to grab a ball but makes an unskilled movement, rolling the ball instead of grabbing it. There are also loss of memories of images or of imagined visual images and problems recalling an object that has a specific structure, etc.

So far, all these were lesions in the dorsal parietal lobe. But if the lesion is closer to the temporal lobe, the deficits become more conceptual and cognitive. The ventral part of the parietal cortex is involved in the more semantic aspects of cognition, like *constructional apraxia* (cannot think spatially), acalculia (cannot do calculations), *agraphia* (cannot write), *alexia* (cannot interpret words). Thus, the lesions that are located lower in the parietal lobe show a transition from space into concepts, as if a lot of our thinking were spatial, and involves both space and ideas.

Attention

We've seen that lesions in the parietal lobe cause both motor deficits and deficits with regard to space and/or abstract space. But another way to interpret the deficits from parietal lesions, like hemineglect contralateral syndromes, is to consider them as deficits in attention, as if the patient was just not attending to that part of the body. But what exactly is attention? Here we have another term that comes from introspection, from our own experience of our mental state, so we should be cautious. For now, let's define attention as perceptual *saliency* by which one particular sensory experience dominates our perception. The same concept of selective filtering can be applied to memories and ideas; it's like there is some sort of searchlight that enables the brain to highlight one thing at the expense of everything else. Attention is critical for survival, since our senses are continuously bombarding us with a stream of information that we have to make sense of.

In fact, attention is deeply connected to novelty, as novel stimuli are automatically detected by the brain while repeated stimuli are ignored. This goes back to the heart of the principle that the brain is a prediction machine: you need to compare what you expect with what is happening and adjust your model of the world to make better prediction of the future. These are all very exciting theories, but remember that they still need to be rigorously demonstrated and anchored in data.

In any case, you could argue that the parietal association cortex is fundamentally involved in attention, and its maps of the world are actually saliency or attention maps, used to highlight one aspect of sensation versus others. In fact, data from monkey recordings indicate that responses of the neurons and the associated maps change dramatically as a function of attention during a behavioral task (figure 18.9). This is as if you were able to zoom in or out of a particular sensory experience, just like you can zoom in and out of a digital map when you want to see more or less detail. So it makes perfect sense that the brain would be correcting its predictions and changing the saliency maps on the fly, inhibiting some parts while enhancing others.

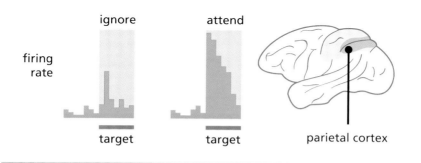

18.9 Neurons in parietal cortex have stronger responses if the animal is **attending** to the stimulus.

There are different types of attention. For example, there is endogenous (internally driven) attention and exogenous attention (e.g., attending to a loud noise). Attention can also be selective to one sensory modality, or supramodal or cross-modal, involving several sensory systems (like attending to the voice of a loved one when you see them). And many parts of the brain are involved in attention, not just the parietal lobe. The frontal eye fields, a motor area in the frontal lobe that controls saccades (which directly reflect our attention); the visual cortex (where lesions lead to *simultanagnosia*, meaning you can't perceive or attend to more than one object at a time); and the pulvinar nucleus of the thalamus are all associated with attentional deficits.

There are different theories of how attention works. Some experiments suggest a *central executive control* frontoparietal network, which jointly decides where to focus the "searchlight." Other data can be explained better by local decision, or *local executive control*, in which individual cortical areas could carry out attentional decisions, biasing signals. Crick argued that there is a specific part of the thalamus that serves as the searchlight. Other models propose that attention is an emergent phenomenon, generated by network-wide attractor states as distributed decision-making without a central command. Finally, some models argue that what we call "attention" is simply the result of reflexive actions based on past experiences.

Understanding the neurobiology of attention can also get tangled with understanding the neurobiology of the self, or consciousness, as attention is always defined in the first person. Fascinating, yet complicated.

Frontal Cortex

Our tour of the association cortex brings us finally to the frontal cortex, and in particular the prefrontal cortex, which is enormously developed in humans. Actually, if you were to try to identify which part of the brain makes us human, you would probably argue that it is the prefrontal cortex. Unfortunately, it is the least understood part of the cortex, so it's a challenge to discuss it. What is clear is that the dorsal and ventral streams both project to the frontal lobe, and, according to Luria's model, you will reach the top of the pyramid, the front of the frontal. The tippy top could be the orbitofrontal cortex (the pre-prefrontal cortex, so to speak), which then projects down to prefrontal areas and then premotor areas. In this simple sketch, the orbitofrontal appears to be involved in emotional labeling, the prefrontal in the cognitive character of movement or intention, whereas the premotor is involved in motor planning.

But the reality is more complicated. In recordings from monkeys, for example, neurons in the frontal association cortex are involved in all kinds of things, like mirroring actions, working memory, and coding expected rewards and plans of actions. Lesions of the frontal association cortex are fascinating, but are also all over the map. Some patients display a striking personality change after the lesion. Phineas Gage, a famous railroad engineer who lost a large part of his frontal lobe due to an accident, became disinhibited and unreliable and lost his sense of direction in life. Other patients with frontal cortex lesions lose their goals; become capricious; get stuck on the first thing that hits their mind, even if it leads to disastrous consequences; are uninhibited; show flat emotion; do not care about the consequences of their actions; and have an overall lack of drive. Often they act as if there is no conscious oversight, as if there is no filter on their behavior, and they can't organize behavior in a way that makes sense. This lack of executive functions, incidentally, also happens in children, perhaps related to the fact that their prefrontal cortex does not fully develop until adolescence. It also occurs in people with schizophrenia, which especially affects the prefrontal cortex. Other prefrontal patients have problems following rules, performing poorly in the Wisconsin test, in which rules of a card game change throughout the test to which they cannot adapt.

Prefrontal patients also have problems in verbal fluency, losing the ability to find the correct word, and they show poor working memory, poor decision-making, distractibility. They perform poorly in gambling tasks, which involve calculating probabilities, often displaying a strong bias toward avoiding a negative outcome, even if it led to greater losses. One could argue that this is all about control, and that the function of the prefrontal cortex—the largest anatomical difference between our brain and that of other animals—is to exercise superb control of our behavior. We can do it, and they can't.

Trying to fit all of this into Luria's simple model just doesn't work. First, there are a lot of cortical areas in the prefrontal lobe (again, it is enormous, which is why our foreheads are so prominent), with many different areas that appear involved in different functions (figure 18.10). Second, if anything is becoming clear from all the patients with lesions it is that the prefrontal cortex, rather than a pinnacle of cognition, essentially provides a negative control on behavior, as if it is putting a lid on a lot of behaviors that need to be controlled and directed because they are inappropriate or ineffective.

And, third, importantly, there does not seem to be a clear hierarchy, but more like a confederacy, a network with many nodes that can come in and out depending on the task. So much for the Luria model! It was a useful way into cognition, but we should now leave it behind.

Let's examine what the prefrontal cortex does in more detail, beginning with its anatomy, as we have often done in the book. The old wisdom that, when the function is unclear, let's approach it from the structure.

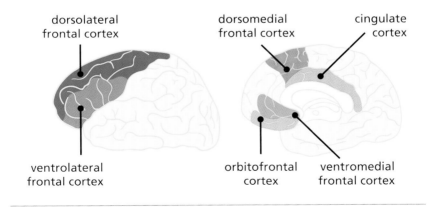

18.10 The **prefrontal cortex** is very large in humans and is involved in higher cognitive functions.

So now we will discuss different parts of the prefrontal cortex, using results from brain imaging, patient lesions, monkey recordings, and microstimulation experiments to try to put together a coherent picture.

Prefrontal Cortex

Let's start with the *premotor cortex*, an old friend from lecture 12 on motor planning. If you record from neurons in the premotor area, you find neurons that are engaged when the animal is both performing an action and receiving sensory information. Looks like those neurons are using some sort of code whereby sensory and motor information is mixed. In order to generate motor commands, the neurons take in sensory information. This could be efficient computationally and goes well with the idea that we are not a circuit but a neuronal network, in which neurons serve many different functions according to the goal of the network. Also, if you think about it, the brain doesn't really care whether something is sensory or motor. So maybe we can consider premotor areas as cognitive, as they are making decisions and implementing plans, ideas, and commands. Is the premotor cortex perhaps the site of our free will? Do we have free will? Time will tell.

Then comes the *dorsolateral prefrontal cortex*, which we've also already met, in lecture 17 on memory. Neurons there are active in working memory tasks, like the delay match to sample. You'll recall that the monkey was looking at a picture that appeared on the screen, which he needed to keep in mind to receive a reward. These dorsolateral prefrontal neurons fire in a sustained fashion until the reward is given, as if they are a type of memory buffer that keeps track of things over short periods of time. This explains why patients with prefrontal lesions lack persistence. It could be that their behavior is due to their forgetting what they are supposed to be doing. Out of sight, out of mind! Admittedly, their symptoms are much longer term than the monkey's short-term working memory cells, which suggests that some neurons in the dorsolateral prefrontal cortex could be coding for both short-term and long-term memory.

As a good neural network, the cortex should have neurons that code for all kinds of situations and intervals. Moreover, working memory is essential for comparisons between different behavioral outcomes and probably underlies behavioral flexibility. Indeed, these patients also have trouble with changing rules in card games.

Let's now move to the *orbitofrontal cortex*, the tip of the pyramid, at the very front of the brain; it is often also grouped together with the *ventromedial frontal cortex*. One way to interpret what goes on in this area is that it computes the value (or credit) of behavior. Let's look at an experiment where a monkey is taught to associate the picture of an object, let's say, a raisin—with a reward (figure 18.11). But, additionally, there is another object to choose from—an apple, for example. Monkeys appear to absolutely love raisins, and therefore this specific neuron in the orbitofrontal cortex fires a lot when a raisin is presented to the animal but not when the apple is shown. Now, let's present the same monkey with a picture of an apple and some cabbage (which everyone appears to hate). But this time, the neuron fires in response to the apple, probably because the monkey prefers apples over cabbage. This is consistent with the hypothesis that the neuron is not coding for the object itself; rather, it is coding more abstractly, for a desired object, for its value. And this also explains well why some prefrontal patients could become flat in terms of emotional drive, because the desire and reward aspect has been taken away from their lives. And how is reward coding implemented? Remember dopamine and the reward prediction error from lecture 13 on motor selection? Well, guess what? The ventral tegmental area—which, along with the substantia nigra, is the main source of dopamine in the brain—projects mostly to the orbitofrontal/ventromedial cortex. So these prefrontal neurons are probably doing sophisticated math, calculating Bayesian probabilities and comparing predictions with outcomes of rewards and updating their chart of values.

The rest of the frontal cortex, which is located closer to the ventral and interior parts of the cortex, is involved in aspects of behavior that have to do more or less with the concept of self. The *ventrolateral prefrontal cortex*, for example, lights up during behaviors that involve self-control,

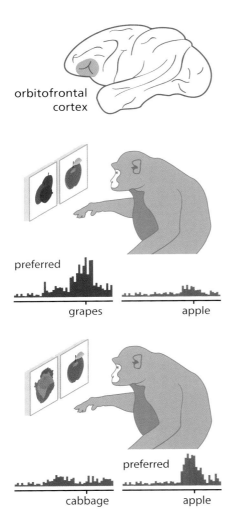

orbitofrontal
cortex

preferred

grapes   apple

preferred

cabbage   apple

18.11 The **orbitofrontal cortex** has neurons that respond to the objects that have higher value for the animal.

when you are trying not to do something, and consistent with this function, is affected in patients with obsessive-compulsive disorders (OCD), depression, and Tourette's syndrome. The *insular cortex*, which we met in lecture 15 on emotions, is indeed active during emotions, during sex and love, and in caring and addictive behaviors. Within the insular cortex, its *anterior cingulate* area becomes active when we make a mistake in our behavior; for example, when behavior is incongruent or conflicting, as if this part of the cortex served as our monitoring "super ego." The anterior cingulate is also affected in OCD and depression and is explored as a target for microstimulation therapies.

Finally, the *posterior cingulate cortex* is particularly interesting; it is affected in schizophrenia, autism, and Alzheimer's and is the main hub of the so-called default mode network during fMRI, which is idling patterns of brain activity that switch on when we are not engaged in behavior but are thinking about ourselves, ruminating, or daydreaming. Is this the site of the self? But what exactly is the self?

Consciousness

We are now close to the end of our journey through the brain, so let's finally tackle consciousness, which is often discussed as being the pinnacle of cognitive ability. What is consciousness? Consciousness is typically associated with self-awareness, but it's not so simple to pin down. It is clear that a lot of our mental activity is unconscious or subconscious, doing things that we are not aware of doing. Unconscious activity is likely the majority of our brain processes. We have met a lot of these processes in the course, from reflexes, to locomotion, to motor learning, to sound identification, etc. As you know, Sigmund Freud thought it was critical that we understand unconscious processing to explain our personality. Maybe. There is an element of personalization in consciousness—your consciousness is subjective and goes with your own sense of self. There is also a unity in the experience of consciousness. You are not conscious of two selves or aware of many things at once. Somehow you know you are one and attend to things, and do things, essentially one at a time.

Consciousness is also associated with intentionality—you are conscious in a purposeful way. There could be perhaps a more primitive or primary consciousness, which we may share with the animal kingdom, and also a human "meta" consciousness, which allows us to be aware of being aware. So, there could be different levels of consciousness that allow for increasing abstraction. These levels of consciousness could fluctuate during the day and when you fall asleep. Clinicians know this well, as patients fluctuate from being fully aware to medically induced coma during, for example, a heart operation. Consciousness is a complicated topic.

Part of the problem with defining consciousness is that it may be another of these introspective terms—and introspection could be completely misleading, a lesson we have learned from what Hubel and Wiesel taught us with the discovery of orientation selectivity as the basis of vision. I bet you consciousness is the same. After more than two thousand years of philosophy on consciousness, with lots of exciting ideas, at the end of the day, we are still waiting for a simple, killer experiment that will illuminate the heart of the problem. This experiment, in my opinion, hasn't yet been done, but we could be getting pretty close.

Let's review what philosophy, psychology, and neuroscience have said about consciousness. As you could predict from the study of an introspective term without strong experimental evidence, there are major disagreements among schools of thought. Not everyone agrees that consciousness is important, though. Perhaps the most radical position, proposed by Daniel Dennet, is that the problem of consciousness doesn't exist as such, and that what we call consciousness is simply a reflection of the brain at work. From this viewpoint, you could look at any part of the brain to find consciousness, because once you "turn on" the brain, it generates your consciousness—it's the natural outcome of brain function.

In the other camp, people like Colin McGinn argue that consciousness does exist but is off-limits to human intelligence, echoing the argument many people make that a machine somehow cannot understand itself. I never quite got this argument.

On the more positive side of the pro-consciousness camp are people like Francis Crick, who, working with Christof Koch, argued that consciousness is something that is generated specifically by the brain and that you should be able to find neurons, or neural correlates, that are active with consciousness. To identify where consciousness is built in the brain, Crick and Koch focused on parts of the brain that are connected to all the other parts of the cortex, one being the thalamus and another one, the claustrum—a tiny part of the brain that is connected to multiple parts of the cortex. Indeed, patients with widespread lesions in the thalamus are rendered unconscious, and the thalamus also seems to switch on and off in the transition between sleep and wakefulness.

In a refined version of Crick's and Koch's hypothesis, some philosophers like John Searle and Thomas Nagel argue that consciousness does exist but is an emergent property that arises from the interaction of neurons, or large groups of neurons, just like magnetism arises from the interaction of particles and not from any individual particle, as you now know so well. So there is no central command or brain area that controls consciousness. Some researchers, like Giulio Tononi, argue that you can actually measure and quantify consciousness with a metric, called "Phi," that mathematically relates to the causal properties of a complex system. Other researchers have argued that consciousness is generated as an emergent property of the interaction between cortical areas. For example, Crick and Wolf Singer argued that a brain-wide 40 Hz electrical oscillation can bind different concepts into a single awareness, forming the essence of consciousness as a distributed oscillating network.

As one example of how consciousness could be an emergent property, in a search for neural correlates of consciousness, Stanislas Dehaene and colleagues performed a simple, elegant experiment. They used fMRI to image the activity of cortical areas that are activated by brief subliminal stimuli (stimuli that subjects were not aware of) and then increased the duration until subjects became aware of them (figure 18.12). What happened was remarkable: subliminal stimuli activated only small, isolated regions of the cortex, mostly in primary sensory areas. But when the stimuli became consciously perceived, they found a massive global reverberating activity throughout the brain, particularly engaging association cortical areas. These results echo the so-called "global workspace model" of Bernard Baars, which proposes that we are conscious of information if it is brought to a central "workspace" in our brain. Thus, consciousness could involve cortical areas more generally and may be related to the default mode network we addressed earlier.

These experiments sound like "Emergent Properties 101," a classic example of a physical phase transition, similar to when the dipoles of a magnet reorient together and generate ferromagnetism. It wouldn't be

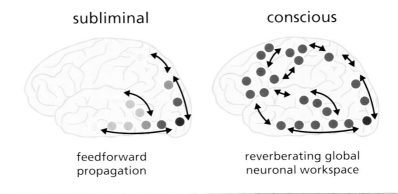

subliminal   conscious

feedforward
propagation

reverberating global
neuronal workspace

18.12 The transition from **subliminal to conscious** stimuli is associated with the widespread activation of non-sensory cortical areas.

the first time basic principles learned from studying simple systems have illuminated the essence of more complicated ones. Science is just beautiful, generating an intertwining fabric of knowledge.

It's one thing to propose a theory; it's another to actually test it. Can one manipulate consciousness experimentally? Couldn't we just switch it on and off by activating or inactivating neurons or brain regions? Easier said than done. Such a research program would be greatly advanced if we could do these types of experiments in animals. While there is a great debate as to whether self-consciousness extends throughout the animal kingdom, I think it is not out of the question to assume that all animals may have a simple concept of self. Why do I say this? Well, you could think of animals as simple automata that react to stimuli with appropriate behaviors selected by evolution. But if this were the case, to generate a new behavior, you would need to wire the whole system over and over again. What a mess! On the other hand, imagine if animals had a concept of self, or a proto-self, generated internally, like an attractor in its neural activity that symbolizes the concept of the animal itself. Then, you could use this proto-self as a building block for any new behavior. In fact, it would be a critical building block for all behavior, since, if you think about it, behavior by necessity must be built to support and protect the life and evolutionary success of the one and only body that generates it. So having the concept of self in your brain could save you a lot of trouble.

In other words, I would argue that without a concept of a self, evolution could only chisel behaviors in a very limiting fashion. But with a concept of self as a tool, evolution would gain the flexibility to greatly expand and explore behavioral outcomes that are productive and phenomenally advantageous, because then the brain could build models of the world in which you yourself are part of the model. In fact, I would take the argument one step further and propose that this is the entire point of the nervous system to implement, with ongoing neural activity, the concept of self. Perhaps the big break that occurred in evolution 750 million years ago, when the first nervous system gave rise to the common ancestors of cnidarians and bilaterians, was the birth of the self, of a primitive form of consciousness. If we can figure this out, we will never look at animals—even humble cnidarians like hydra or corals—the same.

Sleep

After all the introspection and philosophizing, let's come closer to the ground. Let's forget about understanding consciousness for now and consider something simpler, like sleep. To some people, sleep is the absence of consciousness, so maybe this is the key to understanding consciousness. Unfortunately, as you will see, we are quite ignorant about the function of sleep, which is really embarrassing for us neuroscientists, since humans spend about a third of our life sleeping (and even more when we are younger). On top of this, sleep generates a lot of pathologies (who doesn't suffer from insomnia nowadays?), and we still don't really understand why we sleep!

Let's briefly review what we do and don't know about sleep. Sleep is a physiological state generated by the brain. No brain, no sleep. People define it as the lack of wakefulness, but this is a circular definition, since wakefulness is the lack of sleep. In fact, sleep is a precisely orchestrated brain state that is internally generated and accompanied by characteristic

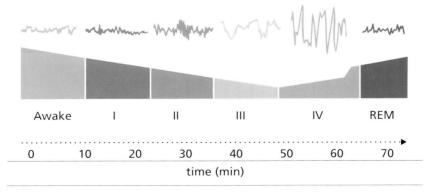

18.13 Sleep progresses through a specific set of stages, characterized by different EEG patterns.

brain rhythms (figure 18.13). Indeed, if we record your EEG while you fall asleep, we find that sleep progresses through a fascinating series of stereotypical stages. When you are awake, your EEG is disorganized, but as you fall asleep, brain waves reflecting the synchronization of neuronal populations in the cortex start to appear. As sleep progresses, the synchronization becomes larger and slower, until you arrive to slow-wave sleep, which is dominated by strong low-frequency oscillations in the EEG. Then, to make it all even more mysterious, you enter the rapid eye movement (REM) stage of sleep, when the EEG resembles the awake state, yet your muscle tone is completely dead, except for your eye muscles, which mysteriously start to make saccades under your closed eyelids. Is this weird or what? Actually not just your eye muscles move, but you also become sexually aroused! Then your brain cycles back and forth through the sleep stages a few times during the night until you wake up, ready for a new day.

How are those stereotyped brain rhythms generated? At the cortical level, the prefrontal cortex is inactivated while activity in the anterior cingulate, amygdala, and hippocampus is enhanced. Large waves of neural synchrony that reach the cortex, generated by subcortical structures, are generating those slow rhythms. Stimulating the thalamus can induce animals to sleep. In fact, the thalamus switches during two modes: when we are awake, the thalamic "gate" is open, enabling the cortical access to sensory information. But when we are asleep, the thalamus generates an internal rhythm that makes the cortex pretty much insensitive to the outside world.

The role of the thalamus is very interesting and perhaps reflects the role the thalamus could play in generating conscious states, as we discussed. But in addition to the thalamus, a variety of brainstem nuclei regulate the transition from sleep to alertness, and stimulating them can put animals to sleep or wake them up. These nuclei include cholinergic nuclei of the reticular formation, raphe nuclei, the locus coeruleus, and the tuberomammillary and VLPO nucleus of the hypothalamus. Indeed, some patients with narcolepsy have specific hypothalamic lesions and deficits.

Though our understanding of how sleep is generated has advanced tremendously, we still are missing the "why." What is the function of sleep? It must be very important, not only because it is precisely orchestrated and it occupies a significant part of our lives and we must spend a lot of energy to generate it and maintain it, but also because it exists in all animals. Even more interestingly, the lack of sleep can have pathological consequences, worsening all kinds of diseases, from dementia to somatic diseases and, if the lack of sleep is complete, can even lead to death!

Yes, if you don't sleep at all, you die. This is not only fascinating but perhaps can give us a clue to what sleep is for. Animals (and people under torture) completely deprived of sleep present a complex clinical syndrome with weight loss, abnormal body temperature, immune deficiencies, memory problems, mood swings, hallucinations, and seizures, all leading to organ failure and death. Looks like a complete systems failure, not just of the brain but of the body. So one conclusion is that sleep affects not just the brain but also the whole body.

Unfortunately, so far the search for the elusive function of sleep has provided more questions than answers. On the brain side, reliable data suggest that sleep enhances the consolidation of memories. But there is more to sleep than that, since you don't need to die from having unconsolidated memories. Crick suggested that sleep and dreams "clean up" parasitic thought processes (whatever that is). Recent exciting data suggest that sleep enhances the internal flow of cerebrospinal fluid, cleaning up the metabolic "debris" from the brain, as if you were flushing the brain of cerebrospinal fluid every night! There is a diversity of opinions. And we haven't even mentioned dreams—no one really understands what dreams are, how they occur, and why we dream.

In terms of the function of sleep for the body, many people have argued that sleep is a general metabolic conservation state that helps replenish glycogen and ATPs, which could explain the organ failure resulting from total sleep deprivation. Evolutionarily, sleep, which occurs mostly during the night, may help avoid activity at night, when it is colder, and increase body activity during the day. I don't know if I buy this. Perhaps this would make sense for visual animals, which need light to operate. Interestingly, species like carnivores (cats) sleep more if they can afford to, so sleep must be desirable (or perhaps plain enjoyable?). All of this only makes the embarrassment of neuroscientists, who still haven't figured out the sleep puzzle, even bigger.

Ending this book on the subject of sleep is a wonderful way to demonstrate the great unknowns we have about the brain. Understanding the brain is of fundamental importance for science, for medicine, and for understanding our minds, which is the ultimate goal of humanism. Yet neuroscience is still young, with many opportunities for research and exploration. It's a fascinating new continent waiting for all of us, waiting for you.

▼ RECAP

Higher cognitive abilities represent the pinnacle of brain function, and represent the core of what makes us human. But how the brain generates them remains a mystery. Most of our cognition appears to involve the association cortex, those parts of the cortex that are not purely sensory or motor. The temporal association cortex builds conceptual maps of the world, while the parietal association cortex generates spatial maps and is also engaged in attention tasks.

The largest part of the association cortex is the prefrontal cortex in the frontal lobe, a fascinating set of cortical areas that appear to be involved in executive functions; that is, planning, selecting, deciding, executing, controlling, and regulating behavior and self-awareness. But allocating specific functions to cortical areas may be misleading if higher cognitive abilities, like consciousness, represent emergent states of brain function that arise from simultaneous interaction between many cortical areas.

Finally, although the phenomenology and mechanisms of sleep are becoming increasingly clear, we still lack a comprehensive understanding of its function, revealing both the nascent state of neuroscience and also the great potentials for discoveries ahead.

Recommended Reading

Baards, B. 1988. *A Cognitive Theory of Consciousness*. Cambridge: Cambridge University Press. The hypothesis of a global workspace model of consciousness.

Dehaene, S. 2014. *Consciousness and the Brain*. New York: Penguin. A review of one of the most interesting proposals on how consciousness is built by the brain.

Koch, C. 2004. *The Quest for Consciousness: A Neurobiological Approach*. Denver: Roberts and Co. Thorough coverage of past and current research on consciousness.

Raichle, M. E. 2015. "The Brain's Default Mode Network." *Annual Review of Neuroscience* 38, no. 1: 433–47. The tale of the discovery, characteristics, and potential function of spontaneous ongoing brain activity.

Xie, L. et al. 2013. "Sleep Drives Metabolite Clearance from the Adult Brain." *Science* 342: 373–77. A beautiful example of how astute observation could give us deep, unexpected insights into the nature of a fundamental question: in this case, the function of sleep.

INDEX

Page numbers in *italics* indicate figures or tables.

nicotinic acetylcholine receptors, 54
NMDA. *See* N-methyl-D-aspartate
NMDA receptor (NMDAR), 41, 57, 60–62,
    246
N-methyl-D-aspartate (NMDA), 41, 55, 67
Nó. *See* Lorente de Nó, Rafael
nociception, 237–38
nociceptive information: descending
    pathway regulation of, 253, *253*, *254*;
    pathways of, 239; processing of, 249
nociceptive stimuli, types of, 239
nociceptive terminals, 239
nonmotor functions, 340
nonmotor learning, 340
nonmotor loops, *307*
nonsteroidal anti-inflammatory drugs
    (NSAIDs), 243
norepinephrine, 346
novelty, 420
novelty detection, 16
nystagmus, 339

object agnosias, 414, *414*
object recognition, 389, 413–14
obsessive-compulsive disorders (OCD),
    426–27
occipital cortex, *412*
occipital lobe, *413*
ocular dominance columns, 151, *153*
oculomotor loop, *307*
odorants, *188*, *191*; combinatorial
    encoding of, 195, *196*; ORs and,
    191–92
odors, 188–89
OFF bipolar cells, 141
OFF-center ganglion cells, 138, *139*
OFF-center receptive fields, *140*
Ogawa, Seiji, 411
O'Keefe, John, 400
olfaction, 189; taste relation to, 199
olfactory bulb, *186*, 189, 190–92, 194–96,
    195
olfactory cilia, 191, *192*
olfactory cortex, *190*, 196–99, *197*, 309

olfactory epithelium, *190*, 191, *191*, 195;
    receptor neurons in, *192*
olfactory hallucinations, 199
olfactory memories, 199
olfactory mucosa, 189
olfactory pathway, *198*, 198–99
olfactory processing, *198*, 198–99
olfactory receptor neurons (ORNs),
    190–93, *192*, 193–94
olfactory receptors (ORs), 191–93, *193*,
    194; combinatorial activation of,
    195, *196*
olfactory system, 188–89, *190*
olfactory transduction, 193, *194*
oligodendrocytes, 52, 70, 210
olivary body, 177, *178–79*, 180
ON bipolar cells, 141
ON-center ganglion cells, 138, *139*
ON-center receptive fields, *140*
On the Origin of Species (Darwin), 130
opiates, 245, 254, *254*
opsins, color vision and, 159
optical microscopy, 92–93
optic apraxia, 419
optic chiasm, 144
optic nerve, 144
optic radiation, 144
optimization, 16–17
optogenetics, 93
orbitofrontal cortex, 308, 423, 425, 426
Organization of Behavior, The (Hebb), 88
OR gate, 108, *109*, 111
orientation columns, 151, *152*
orientation maps, 151
orientation selectivity, 80–83, *82*, 148, 385,
    428; discovery of, 147, *147*; territories
    of, 151
orientation tuning, 83
oriented bars, 147, *147*
oriented receptive fields, 81–82, *82*
OR neuron, 112, *115*
ORNs. *See* olfactory receptor neurons
ORs. *See* olfactory receptors
ossicles, 169

spinal cord, 22, 22, 209–10, 211, 265; axons in, 243, 244; cerebellum and, 330; descending control and, 252–53, 253; dorsal and ventral horns, 244; interconnectivity in, 266; locomotion and, 272; lower motor system and, 261, 261–62; memory and, 399, 399, 400; movement and, 264–66; muscle activation and, 262; pain and, 236; pain pathway and, 240, 243–46; touch pathway and, 222–25; ventral, 266

spinal hemisection, 247

spinal lesions, 246–47

spines, 58–60, 59; dendritic spikes and, 67; excitatory synaptic potential summation by, 66; integration by, 66

spin glasses, 96

spinocerebellum, 329, 329–30, 339

spinohypothalamic tract, 249

spinomesencephalic tract, 249

spinoreticular tract, 249

spinothalamic tract, 249

Spinoza, Baruch, 279, 294

spontaneous activity, 83–84, 87, 117, 122; dynamics and, 121

stapes, 169, 169

STDP. See spike-dependent synaptic plasticity

stereocilia, 172–74, 173–74

steroids, 355

stimulation: motor cortex and, 289–90; spatial patterns of, 215; temporal patterns of, 215

stimulus contrast, 142

stochastic synapses, 90

stochastic transmission, 54, 54–55

stomatogastric ganglion, 85, 85

streams, 132

stress hormones, 358

stressors, emotions as responses to, 344, 359

stress responses, cortisol and, 354

stretch reflexes, 268

striatal neurons, 309, 310, 311

striatum, 306, 309–10; behavior exclusion and, 313; feedback loops and, 307; Huntington's and, 320–21, 321; memory and, 399, 399, 400; Parkinson's and, 319–20; speech and, 374

Structure of Scientific Revolutions, The (Kuhn), 76

Stumpf, Hildegard, 28

subiculum, 395–96

subliminal stimuli, 430

substance P, 243, 245

substantia nigra, 306, 309, 312, 314; inhibitory projections and, 311; Parkinson's and, 319–20, 320

subthalamic nuclei, 306, 316; Huntington's and, 321, 321

sulci, 22

superior colliculus, 143, 180–81, 312

supplementary motor areas, 291

supplementary motor regions, 283

suprachiasmatic nucleus (SCN), 144

surprise minimization, 5–6

symbolic language, 9

sympathetic ganglia, 346

sympathetic system, 342, 346, 347, 348; baroreflex and, 349–50

synapses, 52–55, 53–54; astrocytes and, 70; learning rules and, 105

synaptic cleft, 53

synaptic inputs, 67

synaptic plasticity, 13, 36, 61, 61–63, 89; pattern completion and, 121; Purkinje cells and, 336

synaptic potentials, 66

synaptic refinement, 38–39

synaptic terminals, 55; opiates and, 254; in photoreceptors, 134, 135

synaptic vesicles, 52–53

synaptic weight matrix, 106

synaptic weights, 119; learning rules and, 112

synaptogenesis, 34–36, 35, 60

synchronization, 340

synchronous ensembles, 93

synfire chains, 90–91, *91*

Szilard, Leo, 2

Tank, David, 411

tastands, 200, 205

taste, 199–200

taste buds, 200, *202*

taste cells, 200–201, *202, 204*, 205

taste receptors, 200, *203*, 205

tectorial membrane, 172, *172–73*

telegraph poles wiring, 12

Tello, Francisco, 38

tempering, 55

temporal contrast, 142

temporal cortex, 308, 360, *412, 412–15*;
    speech and, 374, 375

temporal hemiretina, 144

temporal lobe, 359, 372, 396, 399, 409;
    primary auditory cortex in, 181; speech
    and, 374

temporal logic, 26

temporally modulated signals, 182

temporal oscillations, 340

temporal patterns, of stimulation, 215

tendons, 267

thalamus, 23, *178*, 189, 312, 328; attention
    and, 421; auditory, 181–83; basal
    ganglia and, 305; cerebellum and,
    332, *333*; connectivity with, 145;
    consciousness and, 428–29; dorsal
    pathway and, 226; emotional response
    and, *361*; feedback loops and, *307*;
    gustatory pathway and, 200, *201*, 205;
    Huntington's and, *321*; inhibitory
    projections and, 311; nociceptive
    information and, 239; pain pathway
    and, *240*; Parkinson's and, *320*;
    prefrontal loop and, 308; sleep and,
    432–33; somatosensory system and,
    210, *211*; touch pathway and, 225; visual
    pathway and, 80–81, *81*, 131, 143–44

Theory of Colors (Goethe), 165

theory of mind, 295–96

thinking, 408

thoughts, 8; cognitive states and, 386

thyroid, 354

thyrotropin (TSH), 354, 355

thyrotropin-releasing hormone (TRH),
    354, 355

tip link, 172–73, *174*

tissue injury, reactions to, 241–42

tongue, 200, *201–2*, 370

Tononi, Giulio, 429

tonotopic maps, 177

topographic organization, of visual
    pathway, 132

touch, 209–10; channels and receptor cells
    for, 212–19; receptive fields and, *220*,
    220–21

touch afferents, 255, *255*

touch pathway, 224–25; cortical maps and,
    227–28; cortical plasticity and, 230–31;
    descending control and, 232–33; higher
    somatosensory cortex, 231–32

touch receptors, 213, 215, *215–16*

Tourette's, 321–22, 426

transcription factors, 30

transducin, 136

transient receptor potential channels (TRP
    channels), 201, 241, *241*

tree of life, 3

TRH. *See* thyrotropin-releasing hormone

TRIPV1 channel, 241, *241*

TrKA receptors, 246

TRP channels. *See* transient receptor
    potential channels

TSH. *See* thyrotropin

Turing, Alan, 103

TV problem, 94–96, *95*

tympanic membrane (eardrum), 168–69

Ugurbil, Kamil, 411

unilateral reflexes, 258

unimodal association areas, 232

unmyelinated axons, 243

upper motor system, 260, 279–82, *280*

upper temporal lobe, 372

vasoconstriction, 351

vasodilation, 351

vasointestinal peptide (VIP), 346

vasopressin, *355*, 356–57

vegetable agnosias, 157

velocity, 119

ventral horn, 244, *265*; knee-jerk reflex and, 268, *268*; muscle activation and, 262

ventral pathway, 153, *154*, 156–57; speech and, 374

ventral spinal cord, *266*

ventral stream, *412*

ventrolateral frontal cortex, *423*

ventrolateral prefrontal cortex, *425*

ventromedial frontal cortex, *423*, *425*

ventro-posterior medial nucleus (VPM nucleus), 205

verbal working memory, 387, 389

vestibular duct, 171

vestibular nerve, *171*

vestibular nuclei, 329, *331*, 332, *333*

vestibulocerebellum, 329, *329*–30, 339

vibration, 166; measuring, 217–18

VIP. *See* vasointestinal peptide

Virchow, Rudolf, 77

visceral nervous system. *See* enteric nervous system

vision, 130; blind spot, 161–62; color, 157–60; retinal functions, 141–43; retinal processing, 137–41; speed of, 137

visual area MT, 155, *155*

visual cortex, *81*, *131*, 143, *143*, 146–53; attention and, 421; delayed-firing neurons in, 389; neuronal ensembles in, 93; orientation selectivity and, 80–83, 147, *147*; secondary, 150; speech and, 374, *375*

visual pathway, 80–81, *131*, 131–32, 143–46; crossing over in, 144; hierarchical structuring in, 83, 132, *150*, 150–51; visual cortex and, 146–53

visual perception, 384

visual primitives, 153

visual receptive fields, premotor cortex and, 293

visual spectrum, *158*, 159

visual speech, 374

visual stimulus, 14; complex neurons and, 114; neuronal ensembles and, 93, 94

visual system, 80, *81*, 130; anatomy of, 131–32; dorsal and ventral pathways, 153, *154*; interpretation and, 129; as perceptron, 114–17; plasticity and, 337

visuospatial working memory, 387–89; prefrontal cortex and, 388–89

vocal cords, 370, *370*

vocal tract, 370, *370*

volitional movement, 280; cerebellar lesions and, 339

voltage compartmentalization, 60

voluntary movement, 332; cerebellar lesions and, 339

VPM nucleus. *See* ventro-posterior medial nucleus

wavelength, color and, 160

Wernicke, Carl, 371–72

Wernicke's aphasia, 415

Wernicke's area, 371–73, *372–73*, 374, *375*

Wiesel, Torsten, 39, 41, 45, 123, 384; neuron doctrine and, 80; orientation selectivity and, 80–81, 82, 83, 146, 148; perceptrons and, 114, *115*, 116–17; primary visual cortex organization and, 150; visual cortex and, 146

wiring, 10–12, 32–34

wiring diagrams, 10–12

Wisconsin test, 422

working memory, 386; verbal, 387, 389; visuospatial, 387–89

zip code readers, 108–12